A CLASS OF THEIR OWN

HISTORICAL SHIP INTERIORS

BY DEUTSCHE WERKSTÄTTEN

PUBLISHED BY
DEUTSCHE WERKSTÄTTEN

SANDSTEIN VERLAG

A CLASS OF THEIR OWN

HISTORICAL SHIP INTERIORS

BY DEUTSCHE WERKSTÄTTEN

WARSHIPS

OCEAN GIANTS

RICHARD

RIEMERSCHMID

46 **52** **54** **58** **60** **64** **68** **70**

BERLIN

1905

DANZIG

1907

BIGGER,

FASTER,

MORE

EXTRAVAGANT

JOHANN

HEINRICH

BURCHARD

1914

KRONPRINZESSIN

CECILIE

1907

HAMBURG AMERICA LINE

NEW YORK
1927

MAGDALENA
AND
ORINOCO
1928

DEUTSCHLAND
1924

HAMBURG
1926

78 **80** **82** **90** **92** **96** **102** **104**

"MEIN FELD IST DIE WELT!"

ADELBERT
NIEMEYER

KARL BERTSCH

PRESTIGE

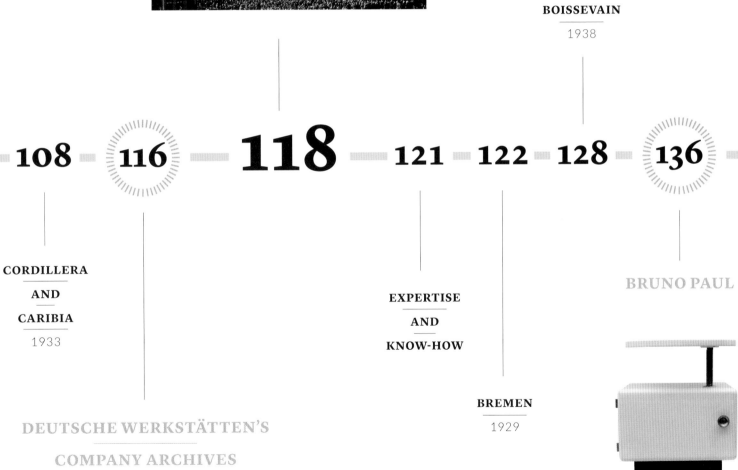

INLAND NAVIGATION

WILHELM

GUSTLOFF

1938

138 **144** **146** **149** **150** **154** **158**

THE BEQUEST OF

WILHELM KRUMBIEGEL

ON LAKES

AND RIVERS

ALLGÄU

1929

LEIPZIG

1929

DEUTSCHLAND

1935

HAMMER AND SICKLE

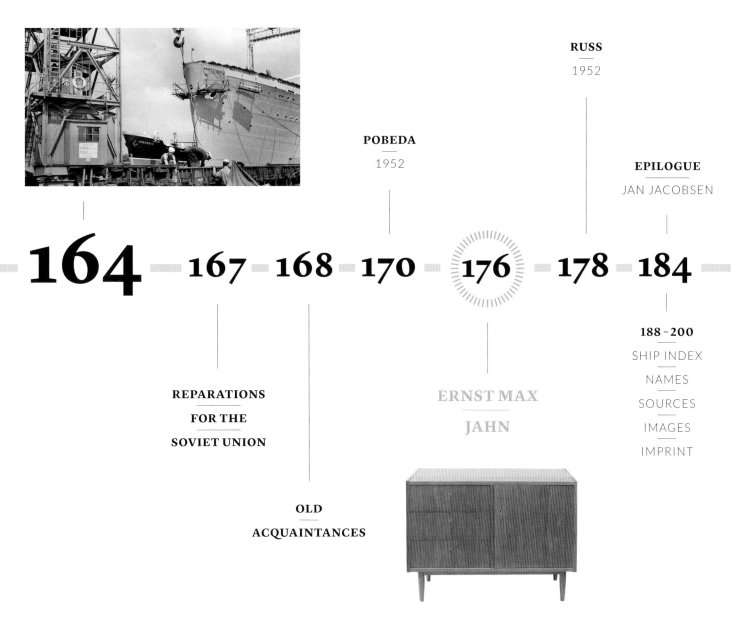

PROLOGUE

Fritz Straub | Managing Partner
Deutsche Werkstätten

Deutsche Werkstätten is always good for a surprise. Back in 2016, we received a request from the world-renowned Victoria & Albert Museum in London, as they began preparing for an exhibition on the design elements of the great ocean liners. We knew, of course, that Deutsche Werkstätten had been involved in furnishing a few passenger steamships in the first half of the twentieth century. But as we took a closer look at the relevant literature as well as a diverse array of art journals from bygone years, we quickly realised we had underestimated the extent and the extraordinary quality of these projects. In the historical archives of Deutsche Werkstätten, which have been housed since 1999 in the Hauptstaatsarchiv Dresden (Dresden State Archive), we also found numerous wonderfully preserved and often beautiful drafts and plans that vividly document the interiors of ships from that era. Probing further, we stumbled upon the collection of a former employee of Deutsche Werkstätten, who had served as a technical supervisor on some of these outfitting projects.

Even before wrapping up our research, we were able to collect a trove of information as well as illustrations and photos. As of this printing, we have confirmed that in the years 1903 to 1907 Deutsche Werkstätten furnished about a dozen warships for the Imperial German Navy. Furthermore, between 1906 and 1938 the company fitted out at least 18 luxury ocean liners, including the *Kronprinzessin Cecilie* (1907), the *Bremen* (1929) and the *Wilhelm Gustloff* (1938). On top of these, Deutsche Werkstätten worked on the interiors of multiple large steamships for inland navigation, one of which, the *Leipzig* (1929), still sails today. Shortly after the Second World War, Deutsche Werkstätten received numerous commissions to refurbish former passenger ships. We have since uncovered five such projects.

Eventually, we circulated some of this information about our history of outfitting ships, including the occasional photo. Time and again, we were met with requests from colleagues, business partners and friends to print 'a little something' on the topic. Here is our answer! And I am very pleased to note that it has turned into more than just a *little* something.

I am equally pleased that Anna Ferrari, from the Victoria & Albert Museum, agreed to provide an introductory text on the international influence of German passenger ships and German design in the first half of the twentieth century.

Her text will transport you back into the dazzling era of the 'ocean giants'. Afterwards we will present – up close and personal – a number of historical examples of Deutsche Werkstätten's craftsmanship from that period.

My thanks also go to Tulga Beyerle, director of the Kunstgewerbemuseum (Museum of Decorative Arts) in Dresden, who has contributed an article on the artistic, cultural and historical significance of Deutsche Werkstätten and its collection. Furthermore, we have received support on this project from the Hauptstaatsarchiv Dresden and the Deutsche Fotothek. I would like to take the opportunity to thank these organisations as well.

And to our readers: I hope you enjoy perusing these pages!

Yours,
Fritz Straub

RIVALRY AT SEA: THE INTERNATIONAL INFLUENCE OF GERMAN PASSENGER LINERS

Anna Ferrari

In 1898, the *Kaiser Wilhelm der Große* created a sensation when it became the fastest liner to cross the Atlantic. Completing the journey in five days and twenty hours, the German liner marked the end of Britain's unrivalled domination of the transatlantic passenger route (Figure 1). Owned by the Bremen shipping company Norddeutscher Lloyd (NDL), the *Kaiser Wilhelm der Große* (1897) was the first German ship to win the Blue Riband, the unofficial and highly prestigious prize awarded to the fastest liner to cross between Europe and North America. The largest and most luxuriously appointed liner yet built, it was also a product of a German shipyard. It was a tangible symbol of Imperial Germany's emergence as an economic and industrial power since the mid-nineteenth century. Those decades of rapid industrialisation witnessed the establishment of two liner companies that became the largest in Germany: the Hamburg-Amerikanische Packetfahrt-Actien-Gesellschaft (known in English as HAPAG or Hamburg America Line) in 1847 and NDL in 1856.

In the last years of the nineteenth century, the launch of the *Kaiser Wilhelm der Große* marked the beginning of an intense international rivalry between the British and German transatlantic shipping companies. Competition extended to the interior design of liners, with shipping lines seeking to surpass each other by offering increasingly luxurious rooms on board. Far more was at stake than just a prize. Capturing the Blue Riband was a question of national pride, as liners became floating national symbols. The idea that liners embodied their country was often made plain in the choice of patriotic names. This was especially true of the *Kaiser Wilhelm der Große* and successive German ships named after members of the imperial family. The largest machines ever built, liners were cast as symbols of national unity and pride for the new German state. Until the outbreak of the Second World War, German liners played a crucial role in the international transatlantic race, spurring a rivalry between companies and nations. HAPAG and NDL initially promoted historicism in the early years of the twentieth century, but they were also among the first shipping lines to invite modern designers to work on their liner interiors.

Kaiser Wilhelm der Große (1897)

Photographed c. 1900

Between 1898 and 1907, the Blue Riband remained in German hands, alternating between NDL and HAPAG. Both companies favoured interiors that harked back to aristocratic homes and appealed to the wealthy international first-class clientele. NDL launched a trio of large liners: the *Kronprinz Wilhelm* (1901), the *Kaiser Wilhelm II* (1903) and the *Kronprinzessin Cecilie* (1906), which featured historicist interiors, densely decorated with heavy gilt mouldings and ornaments. Johann Poppe, the Bremen architect responsible for the *Kaiser Wilhelm der Große*'s interiors (Figure 2), was again employed to design public rooms on board the *Kronprinzessin Cecilie* in a Neo-Baroque style. At the same time, however, NDL also looked to progressive designers and ran a competition for the decoration of thirty first-class cabins on board the ship. The winners of the competition were Richard Riemerschmid, Bruno Paul and Joseph Maria Olbrich, who were soon to become founding members of the Deutscher Werkbund in 1907. This association of artists, designers, industrialists and merchants sought to improve the quality of German applied art and to replace period styles with products suited to modern life and modern German society. Riemerschmid's designs for the Imperial Suite on board the *Kronprinzessin Cecilie* were a stark contrast to Poppe's public rooms and exemplified the new approach to interiors. Panelled in light-coloured wood with a stylised vegetal motif and pared-down furniture, the suite was devoid of historical allusions or excessive, classical ornaments (Figure 3). Riemerschmid's designs were executed by the Dresdner Werkstätten für Handwerkskunst (now

Kaiser Wilhelm der Große (1897)

First-class smoking room

Kronprinzessin Cecilie (1907)

Salon in the Imperial Suite

known as Deutsche Werkstätten) and some of his drawings today survive in the Hauptstaatsarchiv Dresden (Dresden State Archive). The combination of Poppe's interiors with those by Riemerschmid, Paul or Olbrich may seem surprising, but historicist interiors were often highly eclectic, with individual rooms decorated in different period styles. It would not have been uncommon to step from a Louis XIV lounge into an Empire room, for example. Nonetheless, at a time when British shipping lines were firmly wedded to historical styles, this was a significant change in the approach to liner design. NDL grew in confidence, employing Werkbund architects and designers, and appointed Bruno Paul to work on the _George Washington_ (1909). This time, rather than cabins, he designed the first-class public rooms.

By all accounts, these interiors proved successful with the wealthy travellers courted by the most prestigious shipping lines. However, when it came to the design of the _Imperator_ (1913), its most ambitious liner yet, HAPAG adopted a different strategy. In 1909, Albert Ballin, the enterprising director of HAPAG, ordered three giant liners of 50,000 tons each, which were intended to outdo the three British White Star Line sister ships in the planning: the _Olympic_, _Titanic_ and _Britannic_. At 276 metres long and with an imposing superstructure, the _Imperator_'s size allowed for interiors of unprecedented spaciousness. Ballin called upon the most prestigious architect of the time, the Frenchman Charles Mewès, who had conceived the Ritz hotels in Paris, London and Madrid. His interiors for the _Imperator_ effectively recreated a hotel on board. The _Imperator_ was a flurry of historical styles: Louis XIV, Louis XVI, English Tudor and even Roman, with the famous 'Pompeian' columned swimming pool

4

Imperator (1913)

Pompeian-style swimming pool

(Figure 4). It was an astute way of appealing to the travelling elite that had become accustomed to the standards of the new Beaux-Arts hotels springing up across Europe and America. It was a remarkably successful idea, no doubt much envied by Cunard in Britain, which rapidly sought the services of the French architect's British partner, Arthur Joseph Davis, for

Deutschland (1924)

Stateroom sleeping chamber

Deutschland (1924)

Watercolour elevation by

Adelbert Niemeyer

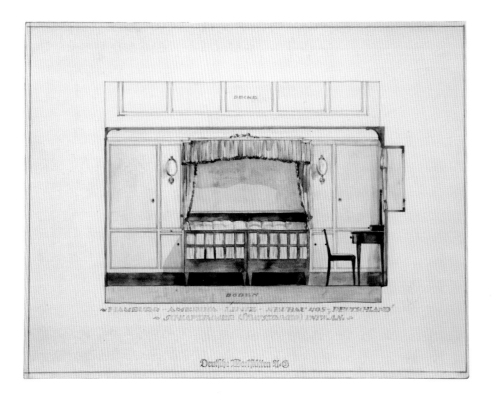

the *Aquitania* (1914). The *Imperator* was not the fastest liner afloat, but together with its sister ship, the *Vaterland* (1914), it established HAPAG as the largest shipping line in the world. On the eve of the First World War, they were powerful symbols of Imperial Germany. Unfortunately, this exceptional German fleet of liners, built in under twenty years, was decimated during the war. Many ships were sunk and those that survived were awarded as reparations to victorious nations. The *Imperator* became Cunard's *Berengaria*, while the *Vaterland* became the American Line's *Leviathan*.

After the war, both HAPAG and NDL rebuilt their fleets. Many interwar German liners were designed in the Art Deco style. This was not merely a question of finding a new identity to represent the young German republic that emerged after the defeat of Imperial Germany, but it was also a savvy business decision following the success of the 1925 Paris Exposition internationale des Arts décoratifs, which showcased the modern streamlined style and gave it its name. The French Line launched two Art Deco liners, the *Paris* in 1921 and the *Ile de France* in 1927, which were each seen as the height of sophistication, luxury and modernity. Despite restrictions, HAPAG built the successful Albert

7

New York (1926)

Second-class ladies' lounge

Ballin class liners, which included the *Albert Ballin* (1923), the *Deutschland* (1924), the *Hamburg* (1926) and the *New York* (1927). The interiors of the last three were produced by Deutsche Werkstätten (Figures 5&6). Although these were not as grand as the pre-war ships, Karl Bertsch's Art Deco marquetry for the *Hamburg* and the *New York* demonstrated the company's embrace of modern styles (Figure 7).

Late in 1926, NDL launched a hugely ambitious programme when it ordered two sister ships for the transatlantic route, which became the *Bremen* (1929) and the *Europa* (1930). Both Blue Riband winners, they were among the most influential liners ever built, setting the tone for the famous French and British Art Deco superliners, the *Normandie* (1935) and the *Queen Mary* (1936), while presenting a new German style on the world stage. Outside, the sisters were streamlined with a rounded stern, a low hull and unusually short funnels, which gave the impression of speed (Figure 8). The interiors differed though: Paul Ludwig Troost designed the first- and second-class interiors on the *Europa* in a neoclassical style, while Fritz August Breuhaus de Groot was responsible for most first-class spaces on board the *Bremen*. Breuhaus de Groot's sleek interiors echoed the modern streamlined silhouette of the liner, as he explained, "the ostentatious luxury of former times, which no longer appeals to the man of today, has been avoided in the interior decoration of the *Bremen* by laying stress on the purity of form, in the beauty of line and on the superior quality of the material."[1] The streamlined interiors were echoed in many later liners, including the elegant HAPAG *Cordillera* (1933). Among the most influential, if not the most visible, features of both sisters were the split smoke uptakes, which took the

Bremen – New York

Poster for Norddeutscher Lloyd, 1930

smoke from the boiler rooms to the funnels without cutting through the centre of the vessel, allowing for uninterrupted vistas down the main suite of public rooms in the centre of the ships (Figure 9). This principle was later exploited to great effect in the enfilade of grand first-class public rooms on board the *Normandie*.

By the end of the Second World War, neither liner remained in German hands: the *Bremen* was gutted by fire in 1941 and the *Europa* was given over to France as reparation in lieu of the *Normandie*, which had burned in New York. Germany's forced disarmament after the war prevented the rebuilding of passenger fleets. In addition, with the extensive development of aircraft and infrastructure, liners eventually became obsolete modes of transport. Nonetheless, companies associated with the heyday of German liners still exist and thrive today. HAPAG and NDL merged to become HAPAG-Lloyd, one of the largest container shipping companies in the world. Blohm & Voss, which built many liners, among which were *Hamburg*, *Cordillera* and *Europa*, still constructs ships. And Deutsche Werkstätten, which produced some of the most forward-looking interiors of German liners, continues to innovate in the outfitting of superyachts.

1 Russell J. Willoughby, *Bremen and Europa: German Speed Queens of the Atlantic* (Surrey: Maritime Publishing Concepts, 2010), 37.

Bremen (1929)
View of the luxurious
shopping arcade

1

Karl Schmidt (front)

Founder of Deutsche Werkstätten

c. 1900

DEUTSCHE WERKSTÄTTEN OR: KARL SCHMIDT, THE VISIONARY FROM SAXONY

Tulga Beyerle

Karl Schmidt was, by birth as well as by training, a simple craftsman – yet at the same time he was a man of extraordinary curiosity, many interests and a great openness to the changing discussion surrounding the decorative arts at the end of the nineteenth century (Figure 1). The journeyman years that followed his carpentry apprenticeship offer proof of his drive to expand his education far beyond the borders of Saxony. The programmatic approach of the Dresdner Werkstätten für Handwerkskunst (founded in 1898, renamed 'Deutschen Werkstätten für Handwerkskunst' in 1907) was influenced by the English Arts and Crafts movement. Schmidt's idea – to work with the best artists in the country, yet still produce affordable furniture – was visionary.

Schmidt had already broken away from the conventional commissioning process by 1899. In contrast to typical practice up to that point, which was to purchase designs from artists (or, in this case, designers and architects), Schmidt instead actively encouraged the professionals – women and men alike – to submit drafts for furniture and handicrafts, offering them a share of five to ten percent of the profit from sales. This demonstrated his astounding feel for the spirit of the times and for the will to reform that was then asserting itself. These reforms included rethinking design, moving away from a historicising mishmash of styles, seeking out new forms and executing them in the highest quality. His proposal of cooperation as well as his network, which was already well activated at that time, produced from the very beginning a multitude of exceptional designs by the era's best designers (Figure 2). In this sense, Karl Schmidt broke new ground in Germany and was soon enjoying international recognition and success. Günther von Pechmann, the first director of the Neue Sammlung in Munich, said the following, and not without reason: "Deutsche Werkstätten is one of the workshops that will always be mentioned when the talk turns to the age of great transformation in German decorative arts [...]."[1] It is therefore no accident that Karl Schmidt also played a decisive role in the founding of the Deutscher Werkbund in 1907.

Through Karl Schmidt's efforts and network, Dresden became a centre for the artistic and technological rejuvenation of interior design and the decorative arts movement. As a businessman, Schmidt understood the value of participating in the most important exhibitions and, accordingly, recruiting the best artists around. However, the innovations touched not only design aspects, but also the technological aspects of furniture production, such as with the development of blockboard. In addition, Schmidt had the farsighted idea of furnishing special showrooms to present a comprehensive range of interior décor. In these showrooms, and in vividly illustrated catalogues, customers could discover not only furniture but also tableware, soft furnishings and wallpaper. It is, therefore, not surprising that Schmidt – with his high standards for quality – began early on to produce wallpaper and textiles as well. The latter of these endeavours led to the 1923 founding of Dewetex (Deutsche Werkstätten Textilgesellschaft; Figure 3). It is interesting that despite its open-mindedness, its innovative courage and its drive towards a modern style of living, the company still sought balance by offering products tailored to the market. Deutsche Werkstätten was never as avant-garde as the Wiener Werkstätte (founded 1903, dissolved 1932) or as radical in its design ambitions as the Bauhaus (founded 1919, forced to disband in 1933 by Nazi

2

Living room furniture

Designed by Josef Maria Olbrich

1903

26

3

Textile sample

Designed by Josef Hillerbrand

c. 1930

repression), but it has always been able to reach a wide audience, and it has survived into the present day despite many crises, nationalisation during the time of the German Democratic Republic and reprivatisation in 1992.

A milestone in Deutsche Werkstätten's history was the introduction of the so-called 'Maschinenmöbel' (machine-manufactured furniture) based on designs presented by Richard Riemerschmid at the Third German Decorative Arts Exhibition in Dresden in 1906 (Figure 4). This was in keeping with the desire to produce high-quality furniture that was still affordable for a large number of people.

4

Linen cupboard

Early machine-manufactured furniture

Designed by Richard Riemerschmid

1906

Deutsche Werkstätten

company buildings

Dresden-Hellerau

c. 1940

At the same time, the company expanded its scope of business and concentrated on the interior furnishing of cafes and other buildings as well as modern modes of transportation, this included passenger ships and railway carriages. Soft furnishings and marquetry work were as much a part of these services as the provision of furniture and wooden ceiling and wall panelling. Another field of activity for Deutsche Werkstätten was the development of prefabricated wooden houses. In this way, Deutsche Werkstätten was able to use its knowledge of and experience with woodworking to help alleviate the housing shortage that followed the First World War. The company developed various models, though it was never able to establish a lasting foothold in the sector.

Karl Schmidt was dedicated to actively helping direct the developments that were taking place in the early twentieth century and participating in the dawn of modernism. So, it is no surprise that he was not satisfied with stopping at the production of furniture, textiles, wallpaper and handicrafts. The early success of his company pushed him to seek out a new production facility. The cooperative purchase of land in today's Hellerau, just outside of Dresden, led not only to the creation of the factory buildings (designed by Richard Riemerschmid, and still standing to this day; Figure 5), but also to the development of one of the first garden cities in Germany (which Riemerschmid likewise played a large role in designing). Influenced partly by Hermann Muthesius, but mostly by the earlier garden cities of England, Karl Schmidt sought to implement the forward-looking idea of working and

Festspielhaus Hellerau

View of the main building

c. 1930

living together in an idyllic space. Workers' estates were built, as were mansions and semi-detached houses. Among the residents of Hellerau were not only employees of Deutsche Werkstätten; artists, artisans, educationalists and members of the literati settled there as well.

Schmidt's far-reaching vision also led to a productive collaboration with Wolf Dohrn, who in turn brought Émile Jaques-Dalcroze, the founder of eurhythmic pedagogy, to Hellerau. There Jaques-Dalcroze staged ground-breaking performances in the iconic Festspielhaus (festival theatre), which was planned by Heinrich Tessenow. The stage itself, designed by Adolphe Appia (with assistance from Alexander von Salzmann), offered just the right setting. In the short intervals between the laying of its cornerstone in 1908, the opening of the Festspielhaus in 1912 and the beginning of the First World War in 1914, the Garden City of Hellerau became one of the centres of avant-garde art in Europe (Figure 6).

Karl Schmidt was and remained an openminded but down-to-earth businessman who was certainly not always enthusiastic about the life-reform activities in Hellerau. Still, that did not stop him from recruiting women as well as men to design his products, or from treating the sexes equally in terms of recognition and payment. Researchers have not yet given Karl Schmidt his due, in terms of both the history of his company and his pan-European professional network. For many years now, historians of post-war German design and architecture have mainly concentrated their research on international modernism. There is an unbroken line of historiography dedicated primarily to the Bauhaus, due to the success of the school's central figures during and following the exile brought on by the Nazis. Other movements coming from Germany and

Europe have received less attention. It would be fitting, in the year before the centennial anniversary of the Bauhaus, to finally give the designers of moderate and decorative modernism the recognition they deserve, people such as Fritz August Breuhaus de Groot, Bruno Paul, Josef Hillerbrand, Else Wenz-Viëtor and Margarete Junge. And we have not yet even touched on the period following the Second World War, when Deutsche Werkstätten was nationalised into a so-called 'Volkseigener Betrieb' (state-owned company) and became the largest furniture manufacturer in East Germany.

Interestingly, the Kunstgewerbemuseum (Museum of Decorative Arts), part of the Staatliche Kunstsammlungen Dresden (Dresden State Art Collections), only began making Deutsche Werkstätten one of the most important focal points of its collection in the 1970s. Thanks to its close cooperation with Deutsche Werkstätten, the museum currently boasts a considerable inventory, one that will continue to be added to in the future. With a collection like this, ongoing research, for us, is just a matter of course. The fact, hardly recognised until today, that about fifty women worked as furniture, textile, wallpaper and handicraft designers has inspired the exhibition 'Gegen die Unsichtbarkeit. Designerinnen der Deutschen Werkstätten Hellerau 1898 bis 1938' (Against Invisibility. Female designers at Deutsche Werkstätten Hellerau from 1898 to 1938). This exhibition, planned for 2018/19, is one of hopefully many upcoming opportunities to explore the history of the company in all its diversity and to enrich the collection of the museum (Figure 7). There is still much work to be done.

1 Günther von Pechmann, 'Die Deutschen Werkstätten für Handwerkskunst GmbH in München', *Dekorative Kunst* 15 (1912), vol. 20, 217–224, at p. 217.

Chair production

View into the old arched girder hall
at Deutsche Werkstätten, c. 1940

WARSHIPS

A SHOW OF FORCE
IN THE AGE
OF IMPERIALISM

OFFICERS' MESSES
AND
CAPTAIN'S CABINS

"HOW I CONQUERED THE NAVY"
(KARL SCHMIDT)

PRINZ ADALBERT (1904)

BERLIN (1905)

RICHARD RIEMERSCHMID

DANZIG (1907)

A SHOW OF FORCE
IN THE AGE OF IMPERIALISM

At the turn of the twentieth century, the German Empire set about building a modern navy. It was spurred on by Emperor Wilhelm II's politically motivated need for recognition, as well as the vision of Alfred von Tirpitz, head of the Imperial Naval Office. In the Imperial Shipyards of Kiel, Wilhelmshaven and Gdansk, workers were now building large numbers of battleships, armoured cruisers, smaller cruisers and even submarines.
A special 'champagne tax' was introduced to pay for the endeavour. This aggressive naval policy had serious consequences, causing great irritation throughout Europe – primarily to the United Kingdom, which, at the time, was the world's undisputed power on the high seas. The British countered by building more warships, ensuring that their naval dominance was at no point seriously challenged. However, the arms race between Germany and Britain damaged the relationship between the two nations. The worsening tensions within the European system of alliances eventually paved the way for the outbreak of the First World War.

Small cruiser

Danzig in an icy roadstead

c. 1910

OFFICERS' MESSES
AND CAPTAIN'S CABINS

Karl Schmidt, the founder of Deutsche Werkstätten (at the time still
known as the Dresdner Werkstätten für Handwerkskunst), was certainly
not lacking in self-confidence. Late in the summer of 1902, Schmidt
travelled to the German Navy in Kiel with a proposition to provide interior
furnishings for their warships. His spirited manner was apparently well
received by the shipyard director, as he was granted an opportunity
to expand on his proposal. Upon receiving this green light, Schmidt left
for Munich, where he met with the well-known artist and architect Richard
Riemerschmid and convinced him to draw up designs for the interiors.
In the years that followed, Deutsche Werkstätten furnished the officers'
messes and captain's cabins on numerous ships. Meanwhile, Riemerschmid
became Deutsche Werkstätten's most important designer, having a lasting
impact on the company's work in furniture production as well.

Dresdner Werkstätten

für Handwerkskunst

Staff photo, 12 June 1901

Dresdener Werkstätten
für
Handwerkskunst.
12.6.1901.

" HOW I CONQUERED THE NAVY "

Karl Schmidt

Karl Schmidt

As a journeyman, c. 1893

"I was about 30 years old and of the opinion that we also needed to be active in the ship furnishing business, and I travelled to Hamburg and Kiel for this purpose. In Kiel I asked to see the shipyard director, Dr. Roßfeld. When I entered the room, I was received with military rigidity by a man with a large black beard. 'What can I do for you?' 'Give me three minutes of your time, your Honour.' 'Go on.' 'When you build ships, you employ the best engineers and designers, and this is right and good! When you furnish the ships, you employ whichever contractor you find, and this is neither right nor good!' 'Oh?!' 'Yes, indeed!' He was a very old man, and my introduction had amused him. I could clearly feel that his heart beat for the modern movement. I got him to the point that I was able to offer to have Richard Riemerschmid draw up plans [...], and one day I returned to Kiel with the drafts and the cost estimates. A large meeting took place. They liked the designs, but they voiced many concerns, since the designs did not conform in this or that way to the shipbuilding code. To which I retorted: 'The designs are excellent; your shipbuilding code is not, it's obsolete. The designs must remain as they are, your code is what needs to change!' Roaring laughter! But the naval office had always been a particularly good public authority – more far-sighted than the others, which is probably related to their profession. The work was done exactly according to Riemerschmid's plans, and it found approval: we went on to furnish the officers' messes and captain's cabins on 12 additional warships, and we became good friends with the gentlemen in Kiel." Deutsche Werkstätten Yearbook 1929

PRINZ ADALBERT (1904)

Imperial Navy | Imperial Shipyard (Kiel) | 126.5 m

Deutsche Werkstätten, following Richard Riemerschmid's designs, fitted out the interiors of the officers' mess on the *Prinz Adalbert*. The ceiling was cladded with sheet brass, in order to withstand the shock that occurred when the ship's guns were fired. The reflective material had the effect of making the room seem less cramped. The fitted and loose furniture also gave the room an appearance more spacious and elegant than was customary for German ships at the time. Riemerschmid's designs, clearly influenced by Art Nouveau, contributed further to this effect. On the day before the cruiser was launched, an excited officer wrote to Riemerschmid:

"The word is getting around, and so not a single day passes without individuals coming to behold this 'attraction' with pleasure [...] As for us officers on board, [we are of] almost a single mind: tasteful, practical and cosy!"

A. Petruschker to R. Riemerschmid on 11 January 1904

The *Prinz Adalbert* was named after the founder and first supreme commander of the Prussian-German Navy, Prince Adalbert of Prussia, initially serving as an artillery training ship. During the First World War, the armoured cruiser was deployed primarily on reconnaissance and convoy missions. After being severely damaged in a torpedo attack in January 1915 and subsequently repaired, the *Prinz Adalbert* was sunk for good by a British submarine off the coast of Liepāja (in present-day Lithuania) on 23 October 1915. From a crew of 675 sailors, only three men could be rescued. The wreckage was first discovered in 2007, at a depth of roughly 80 metres.

Officers' mess

Niche with elaborate partition

RICHARD RIEMERSCHMID   OFFIZIERSMESSE S. M. S. PRINZ ADALBERT
AUSGEFÜHRT VON DEN DRESDENER WERKSTÄTTEN FÜR HANDWERKSKUNST, DRESDEN

Officers' mess

Art Nouveau for the Imperial Navy

Buffet

Unique piece

with opulent brass fittings

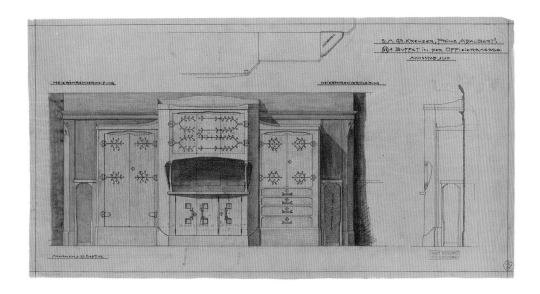

Buffet

Elevation

Richard Riemerschmid, 1902

BERLIN (1905)

Imperial Navy | Imperial Shipyard (Gdansk) | 111.1 m

Before the First World War, the small cruiser *Berlin* was deployed mostly as a reconnaissance ship in the North and Baltic Seas as well as in the Atlantic. She was decommissioned in 1912, but was temporarily brought back into service when the Great War began. The *Berlin* subsequently served as a training ship, among other roles, before being converted into a barracks ship in the 1930s.

Officers' mess

Floor plan

Richard Riemerschmid, 1904

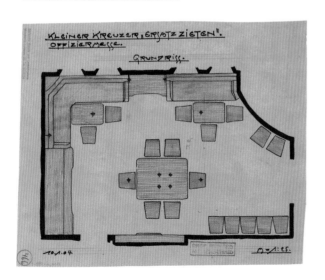

The cruiser was confiscated by the British following the Second World War. In 1947, she was loaded with gas ammunition and scuttled in the Skagerrak.

Deutsche Werkstätten furnished the officers' mess and the captain's cabin aboard the *Berlin*. As he did on the *Prinz Adalbert*, Richard Riemerschmid sought in his designs to create an overall sense of comfort in these spaces, despite the challenging structural limitations of a small cruiser.

Officers' mess
Functional design
in close quarters

Captain's cabin

Fitted buffet

51

Chair 91/8

Oak and leather

Designed 1906

Wardrobe

Pine and plywood with iron fittings

Designed 1908/09

RICHARD RIEMERSCHMID

b. 1868 in Munich | d. 1957 in Munich

Richard Riemerschmid is counted among the most important representatives of the Art Nouveau style. No other designer had such a decisive impact on the development of Deutsche Werkstätten as the artist and architect from Munich. After finishing his studies, Riemerschmid initially worked as a painter. He then turned to designing room interiors, which quickly earned him wide renown. He began his collaboration with Karl Schmidt and the Dresdner Werkstätten für Handwerkskunst in 1902, with the interior outfitting of the Imperial Navy's *Prinz Adalbert*. This was followed by further ship furnishing projects, including numerous designs for pieces of furniture and the complete outfitting of rooms. His 'Maschinenmöbel' (machine-manufactured furniture), first displayed at the Third German Decorative Arts Exhibition in Dresden in 1906, had a lasting impact on interior design in Germany. In addition, Riemerschmid drafted the plans for Deutsche Werkstätten's new factory building as well as for parts of the adjacent Garden City of Hellerau.

DANZIG (1907)

Imperial Navy | Imperial Shipyard (Kiel) | 111.1 m

Like the *Berlin*, the *Danzig* was a small cruiser of the Bremen class and also served as a reconnaissance and training vessel. During the First World War, the *Danzig* was deployed in the North and Baltic Seas. After being hit by a mine and disabled in May 1915, it was the *Berlin*, her sister ship, that towed her in for repairs. Following the war, the *Danzig* was handed over to the United Kingdom, which scrapped her for parts between 1921 and 1923.

Deutsche Werkstätten furnished the *Danzig*'s officers' mess and captain's cabin according to plans drawn up by Richard Riemerschmid. The furniture ensembles of these rooms were displayed at the Third German Decorative Arts Exhibition in Dresden in 1906, where they were highly praised in the trade press:

Captain's cabin

Different perspectives

"When we observe the officers' mess and the captain's cabin on the small cruiser *Danzig*, we are captivated above all by the completely unusual combination of sturdy comfort in the basic forms of the furniture with an almost virtuosic approach to the economic principles of room composition. Here, for the first time, we have an honest, unmistakably German note of dynamic and lively objectivity, which can contest with the international style in nautical cabin design." Dekorative Kunst 9 (1906)

Danzig

Postcard, c. 1910

Officers' mess

With portrait of Kaiser Wilhelm II

OCEAN GIANTS

BIGGER, FASTER,
MORE EXTRAVAGANT

KRONPRINZESSIN CECILIE (1907)

HISTORIC ART JOURNALS

JOHANN HEINRICH BURCHARD
(1914)

Norddeutscher Lloyd Bremen

Poster showing network of routes

1920

Kronprinzessin Cecilie

On a poster for Norddeutscher Lloyd

c. 1910

BIGGER, FASTER, MORE EXTRAVAGANT

The first half of the twentieth century was the age of the great ocean liners. More and more people were deciding to cross the ocean, often for very different reasons: some had the emigrant's dream of a better life abroad, while some travelled simply for pleasure. Those who could afford it travelled first class. Most people, however, had to settle for second class, or even third class (sometimes known as 'steerage'). The route through the North Atlantic was particularly important, and soon large numbers of liners were making the voyage between various European ports and New York. Thanks to improved propulsion technology, cruising speeds on the high seas were continuously increasing. There was also a growing emphasis placed on luxury. A good old-fashioned competition emerged between the large shipping companies, not just over who could transport the most passengers, but also who could build the best ship. This meant not just the fastest or the biggest vessel, but also the one with the most spectacular and elegant furnishings. Some ocean-going steamships had interiors of such opulence that they were reverentially dubbed 'floating palaces'. Due to the tragic sinking of the *Titanic* (1912) and the outbreak of the First World War (1914), the euphoria surrounding ocean liners receded sharply, but it revived during the Roaring Twenties.

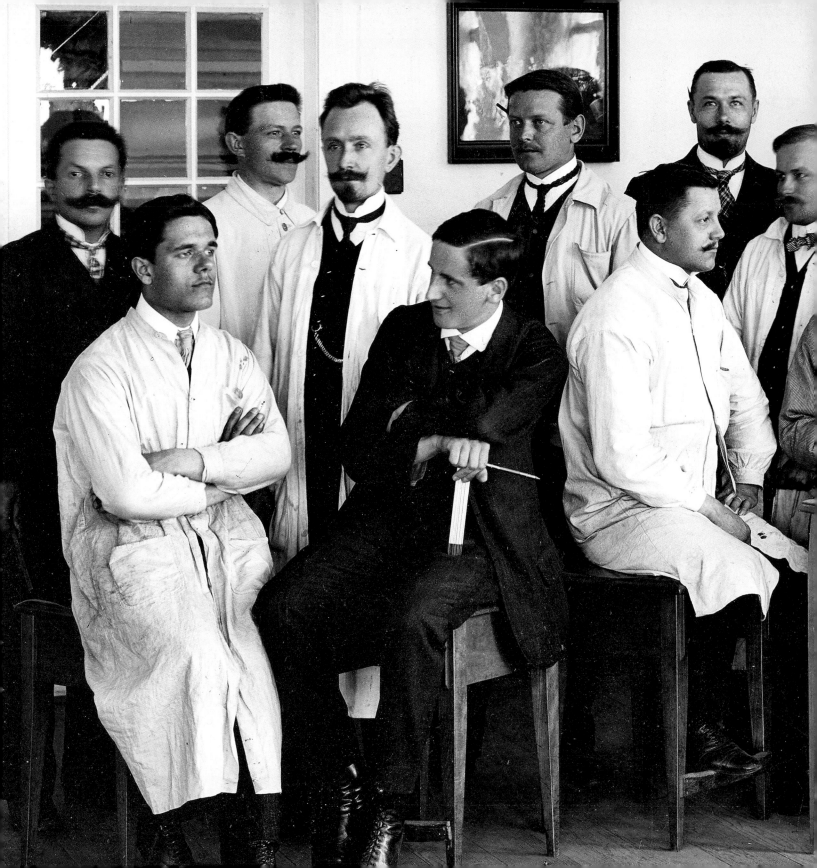

KRONPRINZESSIN CECILIE (1907)

Norddeutscher Lloyd | AG Vulcan (Szczecin) | 215.3 m

The *Kronprinzessin Cecilie* – named after Duchess Cecilie of Mecklenburg-Schwerin, the last crown princess of the German Empire – was one of the fastest, most modern and certainly most beautiful passenger ships of the early twentieth century. Driven by a piston steam engine of 46,000 horsepower, the 215-metre liner could reach speeds of up to 23.6 knots. Like her three sister ships, the *Kronprinzessin Cecilie* sailed for Norddeutscher Lloyd on the North Atlantic route between Bremerhaven and New York. She was commandeered by the United States shortly after the beginning of the First World War and was repurposed in

1917 as the troop transport ship *Mount Vernon*. Once the war ended, the former luxury steamer was decommissioned and left out to rust before finally being scrapped in 1940.

Johann Georg Poppe, Norddeutscher Lloyd's artistic director, was tasked with designing the majority of the interiors on the *Kronprinzessin Cecilie*. In addition, and groundbreaking at the time, modernist interior designers such as Josef Maria Olbrich, Bruno Paul and Richard Riemerschmid were commissioned to plan some of the rooms. Riemerschmid designed the Imperial Suite, which consisted of a breakfast room, a sleeping chamber and a generous salon, all lavishly furnished. Deutsche Werkstätten was again responsible for producing the interiors for this ensemble. Those who desired to travel in such a luxurious suite had to pay between 6,000 and 8,000 marks per crossing – or roughly the price of a small house in Germany at that time.

Imperial Suite

View from the breakfast room

into the salon

Imperial Suite

Sleeping chamber

with ornate ceiling decoration

Salon table

Mahogany and moor oak

The trade press was taken with Riemerschmid's designs and Deutsche Werkstätten's craftsmanship:

"From a sleeping chamber of white and gold the view opens into the salon, which is lent a genteel air by the workings of grey maple wood with enchanting inlays of bloodred rosewood and gleaming white mother-of-pearl in combination with the red upholstery. Connected to the salon is a [breakfast room] consisting of bold forms – with smoky, waxed panelling, wicker chairs and dark-green leather upholstery. [...] the interior design, as a whole, is unsurpassed, displaying the expertise of the old masters." Dekorative Kunst 11 (1908)

Imperial Suite

View of the salon

RICHARD RIEMERSCHMID—MÜNCHEN.   Schlafzimmer in massiv Eiche.

DEUTSCHE WERKSTÄTTEN FÜR HANDWERKS-KUNST
DRESDEN UND MÜNCHEN.

Für den Freund des Kunstgewerbes gibt es kaum etwas Amüsanteres, als einige Stunden in den Verkaufsstellen der Dresdner Werkstätten zu flanieren und zu kramen. Ich weiß nicht zu sagen, wo mir mehr Vergnügen wurde: drüben in den bescheidenen Räumen, die sich die Firma herrichtete, als sie mit der öffentlichen Propaganda energisch einsetzte, oder in den überaus vornehmen Läden, die sie sich vereint mit den Münchner Werkstätten in Berlin baute. Hier wie dort trifft man die gleiche Ware, die gleiche Sachlichkeit und den gleichen Geschmack, der aus trefflichem Material liebenswürdige Stilleben zusammenstellte, trifft man Verkäufer in dem idealen Sinne des Wortes, Fachleute, die einem nichts aufschwätzen, die dem Fragenden Bescheid geben und selbst Bescheid wissen. Man kommt nicht in eines jener unförmlichen Magazine, die mit ihren zwanzig oder fünfzig Musterzimmern renommieren, die einem mit sämtlichen Stilen aufwarten können, die mit derselben Innigkeit ihr Louis seize, ihre Sezession oder die allerletzte Mode preisen, die jeden zivilisierten Menschen nach kurzem Leiden wirblig machen und den Gehrockmann, der tausend unnütze Worte plätscherte, verwünschen lassen. Man kann es kaum anders ausdrücken, man muß sagen: diese Verkaufsstellen haben ihre eigene, wohltemperierte Kultur, sie wirken gepflegt und reserviert und erfreuen durch ihr freimütiges unverhülltes Selbstbewußtsein. Das Prinzip, nach dem sie geleitet werden, ist garnicht zu verkennen: nichts Schlechtes, nichts, was nicht der Zeit und ihrer Art gehört. Wie oft seufzen doch die Geschäftsleute, daß das Moderne nicht ginge, daß das Publikum immer wieder nach dem guten Alten verlange, daß sich nun einmal nicht ändern ließe, man müsse Stil führen, und könne das Neue nur nebenbei protegieren. Das eben ist jene verkehrte Methode, die es aller Welt gerecht machen möchte und dabei nur Unrecht schafft. In den Verkaufsstellen der Dresdner gibt es nicht das, was das Publikum will, vielmehr das, was es haben muß. Dies allerdings in einer so überzeugenden Form und in einer Vollkommenheit, daß selbst arge Skeptiker und träge Gewohn-

HISTORIC ART JOURNALS

German art journals are particularly rich sources for historical information on Deutsche Werkstätten's ship furnishing projects; notably Dekorative Kunst (from 1885), Innendekoration (from 1890) and Deutsche Kunst und Dekoration (from 1897). They document and reflect the radical stylistic changes in German art in the late nineteenth and early twentieth centuries, above all the increasing marginalisation of historicism brought about by the new, fresh forms found in Art Nouveau as well as by the trend towards New Objectivity, both of which had a lasting influence on architecture and interior design. These periodicals are more than just valuable sources of information and graphical documentation – they also provide a fascinating kaleidoscopic insight into the social changes of the times.

JOHANN HEINRICH BURCHARD (1914)

HAPAG | J. C. Tecklenborg (Geestemünde) | 187.4 m

The *Johann Heinrich Burchard* was named after the former mayor of the Hanseatic city of Hamburg, who died shortly after construction began on the ship. She was launched in 1914 by HAPAG – Germany's largest shipping company at the time – but she could not fulfil her planned role on the South American route due to the outbreak of the First World War. She was, therefore, sold to the Dutch company Koninklijke Hollandsche Lloyd while the war was still raging, though she did not leave Bremerhaven until 1920, by which point she had been renamed the *Limburgia*. In 1922, she was

Johann Heinrich Burchard

Launch on 10 October 1914

sold again, this time to United America Lines. Now the *Reliance*, she served first as a liner between Hamburg and New York and later as a cruise ship. Unfortunately, she was badly damaged by a large fire in 1938 and was scrapped three years later.

Due to the criticism of the *Imperator*'s (1913) stylistically bland furnishings, HAPAG reached out to a diverse group of German designers to take on the furnishing of the *Johann Heinrich Burchard*. The influential architect Hermann Muthesius helped HAPAG establish contact with Richard Riemerschmid, Karl Bertsch and Adelbert Niemeyer. Riemerschmid designed the luxury suites on the bridge deck; Niemeyer planned additional suites as well as a ladies' lounge for up to 200 people; and Bertsch was responsible for the staircases. Once again, the designs were manufactured and installed by Deutsche Werkstätten.

To fit out the *Johann Heinrich Burchard*, the designers as well as the interior outfitters had to adapt to special conditions:

"They had to consider the peculiar nature of the ship, whose spaces are shaped by different requirements than those that occur on land. The dimensions are tight; they call for meticulous utilisation. The sloped and curved walls, coupled with the low ceiling, produce a particular sense of space, one which is slightly restricting. Refined proportions must be employed to fool the occupant. Strong profiles and excessively bold ornamentation are to be avoided, as they detract from the room. Anything that is put in real or perceived danger by the ship's movements should be stabilised, not just for its safety but also for its impact on an occupant's impression of the room. When ranging over the walls and their furnishings, the eye should find easy stimulation and minimal resistance." Dekorative Kunst 20 (1917)

Corner sofa

Design drawing

Richard Riemerschmid, 1913

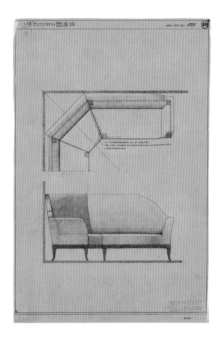

Luxury suite

Views of the living room

Chair

Design drawing

Richard Riemerschmid, 1913

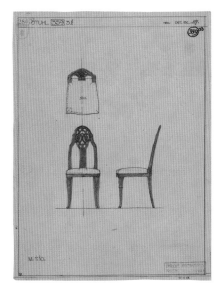

Luxury suite

View of the sleeping chamber

Reliance

Formerly the *Johann Heinrich Burchard*

c. 1937

Staircase

Ascent to the upper deck

HAMBURG AMERICA LINE

"MEIN FELD IST DIE WELT!"
DEUTSCHLAND (1924)
ADELBERT NIEMEYER
HAMBURG (1926)
NEW YORK (1927)
KARL BERTSCH
MAGDALENA AND ORINOCO (1928)
CORDILLERA AND CARIBIA (1933)
DEUTSCHE WERKSTÄTTEN'S COMPANY ARCHIVES

"MEIN FELD IST DIE WELT!"

With this slogan (which translates as 'My Realm is the World'), the Hamburg-Amerikanische Packetfahrt-Actien-Gesellschaft, founded in 1847, developed into one of the largest and most successful shipping companies in the world at the beginning of the twentieth century. Its transatlantic service did particularly well, due primarily to the masses of emigrants leaving for the United States. Besides its ocean liner service, HAPAG, or Hamburg America Line (as it is commonly known in English), was also the first company to offer cruises on a large scale. The First World War, however, put this success story on temporary hold. Numerous HAPAG ships were interned in neutral ports at the beginning of the war. After the fighting ended, the company was forced to turn over almost the entirety of its remaining fleet to the victorious Allies.

HAPAG was able to reattain its levels of pre-war success remarkably quickly after 1918, thanks to government subsidies and its own shrewd cooperation with a diverse group of foreign and domestic shipping companies. By the early 1920s, the firm was again placing construction orders for increasingly modern passenger ships, primarily with Blohm & Voss in Hamburg. HAPAG rapidly built up a new and considerably large fleet, before the global depression, which began in 1929, ground the company to a halt once again.

DEUTSCHLAND (1924)

HAPAG | Blohm & Voss (Hamburg) | 191.2 m

The *Deutschland*, with her two funnels and four masts, had the space to carry more than 1,500 passengers – 198 in first class, 400 in second class and 935 in third class – as she sailed the North Atlantic route for HAPAG. During the Second World War she was repurposed multiple times, first as a barracks ship and later as a hospital ship. She also transported refugees and wounded soldiers, before being sunk in the Bay of Lübeck by British fighter planes on 3 May 1945. The wreckage was raised three years later and subsequently scrapped.

The *Deutschland* marked Deutsche Werkstätten's first collaboration with Blohm & Voss, the renowned shipyard in Hamburg. The commission included fitting out the so-called 'staterooms', which consisted of a lounge and a bedroom, based on plans developed by Bruno Paul and Adelbert Niemeyer. Niemeyer also designed the ladies' writing room; these interiors were likewise fabricated by Deutsche Werkstätten.

Stateroom

Floor plan and elevations of walls

Bruno Paul, 1923

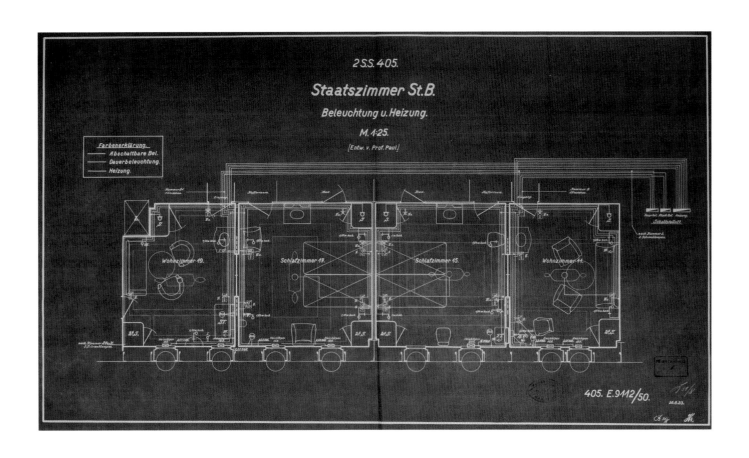

2 S.S. 405.

Staatszimmer St.B.

Beleuchtung u. Heizung.

M. 1:25.

(Entw. v. Prof. Paul.)

405. E.9112/50.

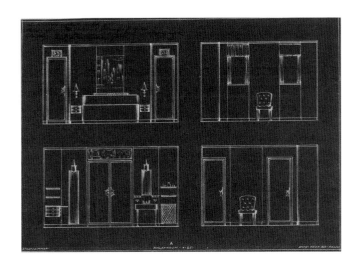

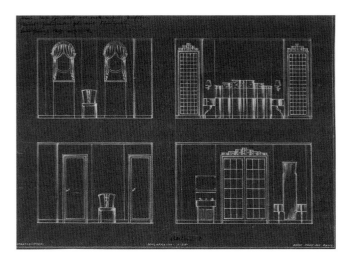

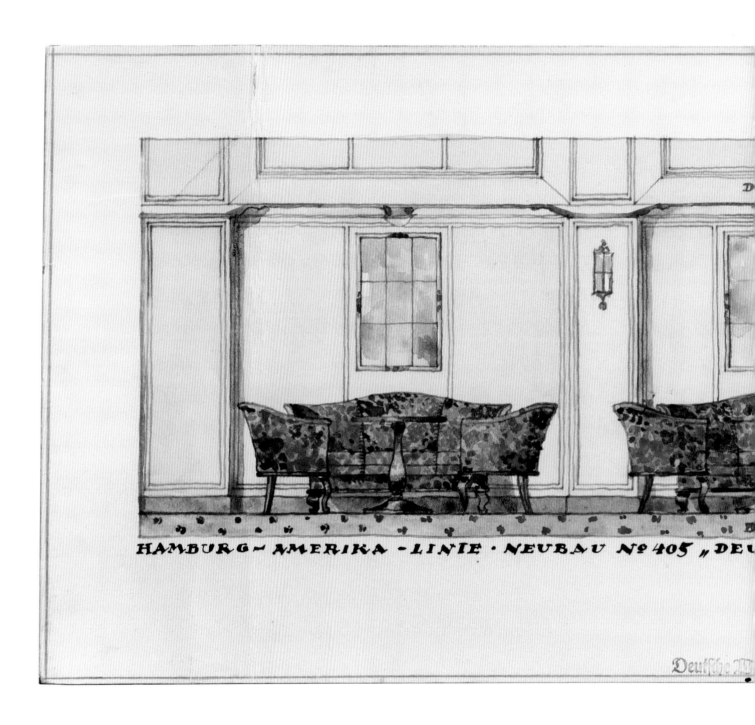

HAMBURG~AMERIKA·LINIE·NEUBAU Nº 405 „DEU

HLAND · DAMENSCHREIBZIMMER · ENTWURF · PROF. AN 1922

Ladies' writing room

Watercolour elevation

Adelbert Niemeyer, 1922

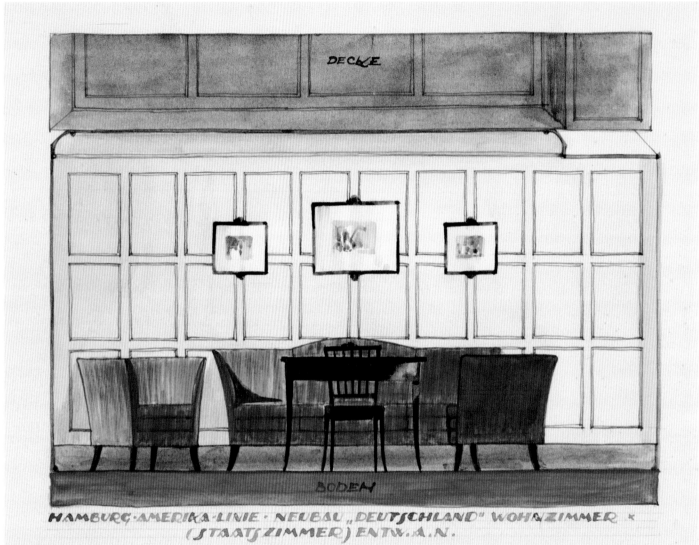

DECKE

BODEN

HAMBURG · AMERIKA · LINIE · NEUBAU „DEUTSCHLAND" WOHNZIMMER ×
(STAATSZIMMER) ENTW. A. N.

Stateroom

Watercolour elevation

of the living room

Adelbert Niemeyer, 1922

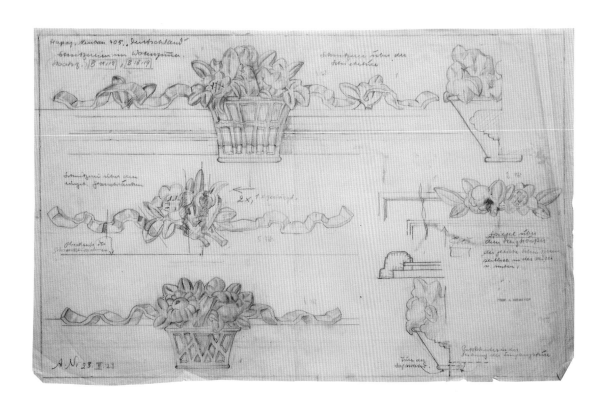

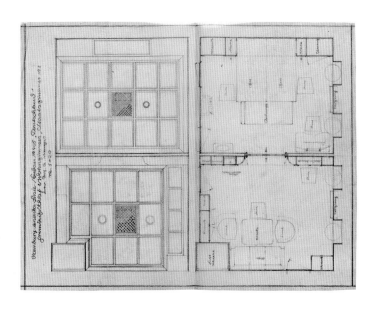

Stateroom

Detail drawing and floor plan

Adelbert Niemeyer, 1922

Living room

c. 1910

Credenza

Cherry wood and mahogany

with brass handles

Designed c. 1909

ADELBERT NIEMEYER

b. 1867 in Warburg | d. 1932 in Munich

Adelbert Niemeyer studied painting at the art academy in Düsseldorf. He relocated to Munich in 1888, where he put his wide range of artistic talents to good use, working intermittently as an actor and musician. In 1898 he was one of the initiators of the Munich Secession art movement. Four years later, together with Karl Bertsch, he founded the Münchner Werkstätten für Wohnungseinrichtung. This company merged with Dresdner Werkstätten für Handwerkskunst (which would eventually rename itself Deutsche Werkstätten für Handwerkskunst), and Adelbert Niemeyer became one of its most significant designers. He devised numerous individual pieces and lines of furniture, as well as wallpapers, fabrics and carpets, even prefabricated wooden houses. He was also involved with furnishing projects on the HAPAG ocean liners *Johann Heinrich Burchard* and *Deutschland*. In addition to his professional activities as a designer, he served as a professor at Munich's college for decorative arts.

HAMBURG (1926)

HAPAG | Blohm & Voss (Hamburg) | 193.5 m

The *Hamburg*, similar to the *Deutschland*, was a passenger ship of the so-called 'Albert Ballin class', likewise sailing the North Atlantic route. She was later employed repeatedly and with great success as a cruise liner. During the Second World War the *Hamburg* operated as a transport and barracks ship, until an exploding mine caused her to capsize in March of 1945. When the war was over, the wreckage was refloated by the Soviet Union and repaired in various shipyards. The Soviets originally planned to turn her into a passenger ship, but in the end she became the *Yuri Dolgorukiy*, a mothership for whaling vessels. The former HAPAG steam liner served in this role in the Southern Ocean from 1960 to 1976.

Second-class ladies' lounge
Finished in flame birch wood

Deutsche Werkstätten furnished the ladies' lounge for the *Hamburg*'s second-class passengers according to plans from Karl Bertsch. The work once again earned many compliments:

"The ladies' lounge in second class – finished in flame birch wood and manufactured by Deutsche Werkstätten from Dresden-Hellerau – is so wonderful in its simplicity that one can practically say no first-class lounge could match it, much less surpass it."

Dresdner Neueste Nachrichten 191 (17 August 1926)

Second-class ladies' lounge

Wall panelling with

intricate veneer work

NEW YORK (1927)

HAPAG | Blohm & Voss (Hamburg) | 193.5 m

On the *New York*, Deutsche Werkstätten fitted out not only the ladies' lounge – as they did on the Hamburg – but also the second-class dining hall. The quality of the work on the *New York*, carried out according to plans by Karl Bertsch, was widely praised:

"Accomplished in style and comforting effect, the interiors of the second class, which were executed entirely by Deutsche Werkstätten – with their wonderfully grained, warm walnut panelling on the tables and walls – are among the cosiest that today's interior design has achieved."

Münchner Neueste Nachrichten 91 (3 April 1927)

The *New York*, built one year after the *Hamburg*, was practically identical to her sister in exterior construction and interior design. Like the *Hamburg*, she served as an ocean liner and a cruise ship, then had duties as a barracks vessel during the Second World War. In 1945, she was sunk off the coast of Kiel by American aerial bombardment. Her wreckage was later raised and towed to the United Kingdom to be scrapped.

New York

Workers painting the bow

c. 1925

Second-class ladies' lounge

Interior design in Art Deco style

Second-class dining hall

Simple, functional, yet still elegant

Dining room 120

c. 1927

Sideboard 120/5

Zebrawood and Oregon pine

with brass top piece

Designed 1927

KARL BERTSCH

b. 1873 in Munich | d. 1933 in Bad Nauheim

Karl Bertsch never studied at any art school – his knowledge of the designer's craft was entirely self-taught. Nevertheless, he became one of Germany's most significant representatives of the modern decorative arts. In 1902, Bertsch, along with Willy von Beckerath and Adelbert Niemeyer, founded the Münchner Werkstätten für Wohnungseinrichtung, which merged with Karl Schmidt's Dresdner Werkstätten für Handwerkskunst in 1907. From then on, the company did business under the name Deutsche Werkstätten für Handwerkskunst. Bertsch held a managerial role from the very beginning, though he simultaneously carried out a range of design tasks, including the creation of hundreds of individual pieces of furniture as well as whole room interiors. He was also Deutsche Werkstätten's most important designer for ship outfitting projects during the 1920s, for ocean-going as well as inland navigation vessels.

MAGDALENA
AND ORINOCO (1928)

HAPAG | F. Schichau (Gdansk) + Bremer Vulkan (Bremen) | 147.5 m

The sister ships *Magdalena* and *Orinoco* were powered by two large 6,800 hp diesel engines each. Starting in 1928, they sailed to Central America and the West Indies for HAPAG. This did not last long, however, as fate had different plans in store for both ships. The *Magdalena* ran aground near Curaçao at the beginning of 1934 and had to be towed off and overhauled, but she did sail again, under the name *Iberia*. Following the Second World War, she was given to the Soviet Union as reparations. The Soviets renamed the ship *Pobeda*

104

First-class social hall

Photographed on the *Magdalena*

(Victory) and deployed her as a steam liner and cruise ship on the Black Sea. The *Orinoco*, on the other hand, was confiscated by Mexico in 1941. She was initially loaned to the United States and was later sold to Argentina. In 1947, under the name *Juan de Garay* and with only one class of accommodation, she began travelling back and forth between Río de la Plata and the Mediterranean ports of Europe.

Deutsche Werkstätten furnished the first-class social hall on both the *Magdalena* and the *Orinoco* in the elegant style of 1920s chic. Particularly striking was the large ceiling luminaire, which bathed the room in a pleasantly diffuse light. Karl Bertsch supplied the interior designs.

First-class social hall

Views on board the *Orinoco*

CORDILLERA
AND CARIBIA (1933)

HAPAG | Blohm & Voss (Hamburg) | 159.8 m

Construction of the *Cordillera* and the *Caribia* saved the large Hamburg shipyard Blohm & Voss from bankruptcy during the global depression of the early 1930s, along with financial support from the government. Starting in 1933, the two sisters sailed the Central American route on behalf of HAPAG, also making port in Caribbean islands. During the Second World War the steamships were temporarily used for military purposes. Afterwards they went as spoils of war to the Soviet Union, which refitted them to their original purpose as passenger liners. The *Cordillera* was renamed *Russ*, while the *Caribia* became the *Iljitsch*.

A special feature of the sister ships was the open central axis, which was made possible by placing the exhaust shafts at the sides rather than centrally (as was customary). Deutsche Werkstätten fitted out the opulent first-class dining hall on both the *Cordillera* and the *Caribia*, yet again following designs by Karl Bertsch. The craftsmanship constituted a complete work of art:

"Due to the selection of wood types (birch and walnut), the white lacquered ceiling and the colour-coordinated carpets, upholstery and drapery (produced by Deutsche Werkstätten), the room appears select and harmonious with regard to the colours and materials. [...] The ship is a valuable testimony to German creative abilities, on the part of both the artist who designed it as well as the company that executed it." Innendekoration 44 (1933)

First-class dining hall

Thoroughly composed comfort

on the *Cordillera*

Cordillera

Launch on 6 March 1933

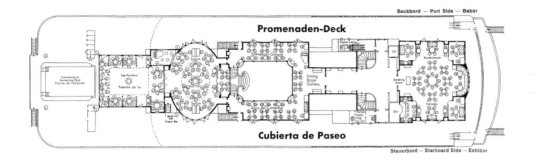

Promenaden-Deck

Backbord — Port Side — Babor

Steuerbord — Starboard Side — Estribor

Cubierta de Paseo

Deck plan of the *Caribia*

c. 1932

Caribia

c. 1935

First-class dining hall

Impressions from the *Cordillera*

Wall decoration

Design drawing for the *Cordillera*

c. 1932

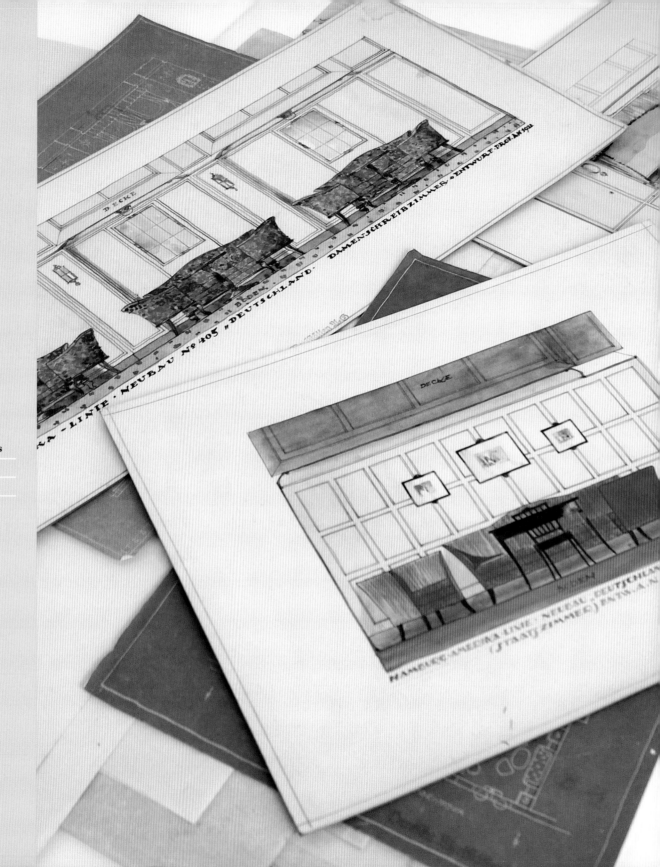

Folder with design drawings

For the HAPAG steamer

Deutschland

DEUTSCHE WERKSTÄTTEN'S
COMPANY ARCHIVES

During the first half of the twentieth century, Deutsche Werkstätten had a lasting impact on interior design and design in general in Germany. The company revolutionised not just furniture production but all aspects of interior outfitting. It collaborated with the era's most renowned architects and played a leading role in the founding of the Deutscher Werkbund (1907) and the Garden City of Hellerau (1909–1912). Luckily, much of this is well documented, since Deutsche Werkstätten, from the very beginning, preserved photos, drafts, plans, letters and other papers, including copious materials related to ship furnishing projects. In 1999, the entirety of the company's archives was transferred to the State Archive in Dresden. The documents are not only stored there under optimal conditions but are also readily available for researchers and others who are interested in viewing them. Covering over 250 metres of shelf space, the inventory with call number 11764 is a legally protected cultural asset.

PRESTIGE

Installation work

Deutsche Werkstätten employees
on the *Wilhelm Gustloff*, c. 1937

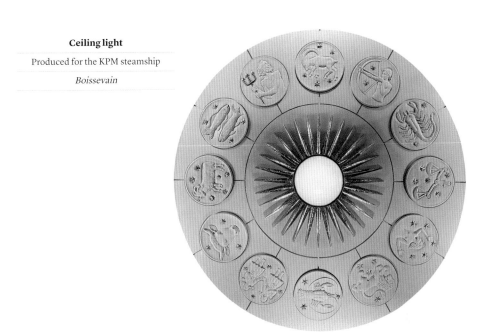

Ceiling light

Produced for the KPM steamship

Boissevain

EXPERTISE AND KNOW-HOW

Deutsche Werkstätten was Germany's largest and most successful furniture
manufacturer in the first half of the twentieth century. The company's extensive range
could be viewed and ordered from their own catalogues and outlets in most major cities.
Even in those days, however, there were company divisions dedicated to special projects.
Deutsche Werkstätten fit out large numbers of public buildings as well as privately
owned residences and, for a time, even built prefabricated wooden houses, some of
which were sold abroad. Ship outfitting projects were a limited yet exceptionally
prominent part of such activities. Deutsche Werkstätten was seen as a reliable partner,
and it enjoyed an outstanding reputation among the shipyards and shipping firms.
When fitting out ocean liners, the company was able to draw on its experiences with
interior furnishing and furniture production, experiences that provided comprehensive
knowledge of materials as well as a diverse range of self-developed manufacturing
technologies.

BREMEN (1929)

Norddeutscher Lloyd | DeSchiMAG (Bremen) | 286.1 m

Besides HAPAG, there was another extremely successful shipping company in Germany: Norddeutscher Lloyd, founded in 1857 and based in Bremen. NDL (as it was known) caused a sensation at the end of the 1920s with two huge projects: the *Bremen* and her sister ship the *Europa*. Their construction alone received tremendous attention from all over the world. On her maiden voyage to New York, the *Bremen* broke the record for fastest transatlantic crossing and was awarded the prestigious Blue Riband. The 286-metre-long, 4-screw express steamer was not only technically impressive, but its interiors, mostly designed with first- and second-class passengers in mind, were imposing, too. During the Second World War, the *Bremen* was painted grey and refitted as a troop transport ship. On 16 March 1941, while she was docked in the Port of Hamburg, a large fire broke out on board and she had to be flooded. All usable parts were stripped from the vessel, which was subsequently scrapped.

The renowned architect Fritz August Breuhaus de Groot was contracted to design the *Bremen*'s public rooms. However, many additional designers, along with the best outfitting companies in the country, also provided interior furnishings for what was, at the time, the world's most cutting-edge passenger ship. Deutsche Werkstätten fit out a luxury suite based on plans by Bruno Paul. The elegantly restrained ensemble consisted of a bedroom and a living room, both panelled with bright satinwood, as well as a hallway and a bathroom. Cabinets, shelves and a range of drawers were exquisitely integrated into the walls.

Luxury cabin

View of the living room

Bremen

Launch on 16 August 1928

Norddeutscher Lloyd

Bremen.

Weihnachten 1929.

ur bleibenden Erinnerung an die Indienststellung
unseres Schnelldampfers „Bremen" haben wir eine
Medaille prägen lassen, die wir denjenigen Mit-
arbeitern am Bau dieses Schiffes zu verleihen
beschlossen haben, die sich hierbei besonders
hervorgetan haben. — In Anerkennung Ihrer
verdienstvollen Mitarbeit gestatten wir uns Ihnen
beifolgend eine solche Medaille zu überreichen,
die Ihnen eine dauernde Erinnerung an diese
Großtat deutschen Könnens bleiben möge.

Norddeutscher Lloyd

Herrn Wilhelm Krumbiegel

Klotzsche

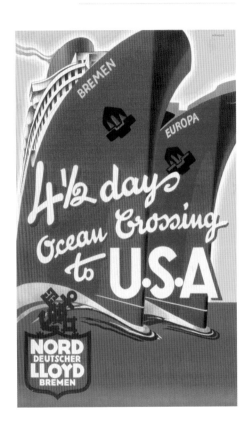

Bremen and _Europa_

Poster for Norddeutscher Lloyd

c. 1932

Bremen

In New York Harbor

1938

BOISSEVAIN (1938)

Koninklijke Paketvaart Maatschappij | Blohm & Voss (Hamburg) | 170.5 m

The *Boissevain* was built in Hamburg by Blohm & Voss, on commission from the Amsterdam-based shipping company Koninklijke Paketvaart Maatschappij. She was the firm's first ship built in Germany. But this was not the only unique thing about her: the shipyard was paid for its work not in money, but instead in a huge quantity of tobacco (colonial goods). Due to her appearance, the *Boissevain* was also known as the 'white Dutch yacht'. Following her launch in 1938, she operated primarily in the waters off modern-day Indonesia. The *Boissevain* was transferred in 1948 to the Koninklijke

Java-China-Paketvaart Lijnen, for which she sailed the so-called 'Asia-Africa-South America line'. In 1968, after 30 years of service, she was retired and dismantled.

Deutsche Werkstätten was particularly proud of its craftsmanship on the *Boissevain*. Due to its good reputation, the company was hired to fit out a large portion of the 170.5-metre-long East Indies steamer:

"The shipping company awarded the contract for the foyer, lounge, dining hall and staircases to Deutsche Werkstätten Hellerau, a firm with much experience fitting out ships and with modern production facilities."

Innendekoration 49 (1938)

Foyer

Gallery area with decorative maps

Lounge

View of the smoking room

In fitting out the *Boissevain*, Deutsche Werkstätten made extensive use of so-called 'TEGO plywood', with layers compressed at high pressure without the aid of glue. This made the boards resistant to moisture as well as to drastic changes in temperature. The plans for the interiors were supplied by one of Germany's most highly regarded designers in the ship-furnishing industry: Bruno Paul.

Dining hall

With 'Sumatra-Java-Borneo'

tapestry

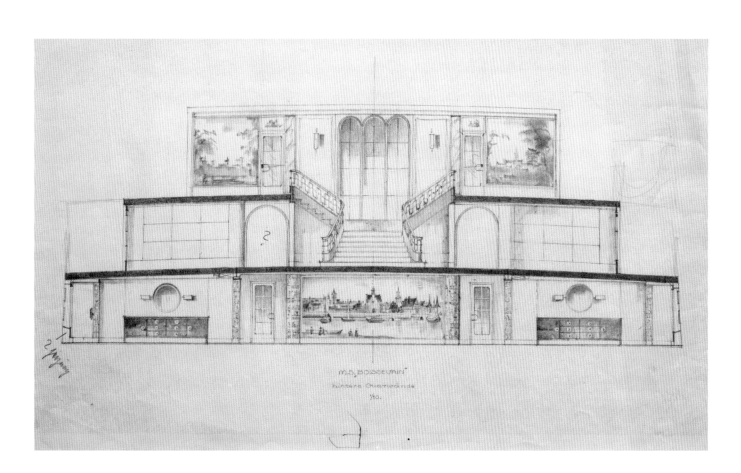

M.S. BOISSEVAIN

Elevation of walls

c. 1937

Smoking room

With booth

Nightstand 273/10

Oak and pine, lacquered yellow

Designed 1932

BRUNO PAUL

b. 1874 in Seifhennersdorf | d. 1968 in Berlin

As one of the best-known and most influential designers of his day, Bruno Paul was a pioneer in modern architecture in Germany. Following his studies at Dresden's college of decorative arts and later at the Academy of Fine Arts in Munich, Paul worked briefly as an illustrator and caricaturist. However, he soon directed his efforts towards architecture and interior design. He was entrusted with his first project for Deutsche Werkstätten in 1911. Throughout the 1920s and 1930s, he was one of the company's most active and important designers. During this time, he strongly influenced Deutsche Werkstätten's mass-produced furniture lines, but he did not stop there. He also designed luxurious room interiors, prefabricated wooden houses and interior furnishings for the ocean liners *Bremen* and *Boissevain*. Aside from these endeavours, Paul also found tremendous success as an architect and as a college instructor in Berlin, where Ludwig Mies van der Rohe was one of his students.

WILHELM GUSTLOFF (1938)

Deutsche Arbeitsfront | Blohm & Voss (Hamburg) | 208.5 m

The history of the *Wilhelm Gustloff* is as problematic as it is tragic. The vessel was built as a cruise ship commissioned by the Nazis' so-called 'Deutsche Arbeitsfront' (German Labour Front), specifically for its recreational division 'Kraft durch Freude' (Strength through Joy). State-of-the-art furnishings were provided for the almost 1,500 passengers. All cabins had an exterior window, and there was a sports area on the sun deck. The *Wilhelm Gustloff* followed up her maiden voyage to Madeira (a Portuguese territory) with several trips to Norway and Italy.

During the Second World War, the *Wilhelm Gustloff* was initially refitted as a hospital ship and later as a troop transporter. On the night of 30 January 1945, she was hit by torpedoes from a Soviet submarine patrolling the Baltic Sea, subsequently sinking not far from the coast. The *Wilhelm Gustloff* was over capacity at the time, mostly with refugees leaving Farther Pomerania (now Poland). Though the number of people on board was never exactly determined, estimates range up to 10,300 passengers. In the end, just over one thousand could be rescued.

Great Hall

Installation work, c. 1937

Great Hall

Installation work, c. 1937

The architect Woldemar Brinkmann was tasked with designing the interiors of the *Wilhelm Gustloff*. Deutsche Werkstätten furnished the cruise ship's so-called 'Great Hall' which accomodated two dance floors and up to 380 passengers on the lower promenade deck.

The Great Hall was almost entirely executed in light-coloured high-gloss lacquer. It contained a music podium, and numerous large paintings were hung on its walls. The complete outfitting of the hall – including panelling and cladding, loose furniture, decoration work, integration of technical elements and other aspects – was directed and coordinated on site by Deutsche Werkstätten employees.

Great Hall

Impressions

Wilhelm Gustloff

Shortly before being put into service

March 1938

New York

During construction by Blohm & Voss

in Hamburg, c. 1925

Wilhelm Gustloff

In the Port of Hamburg

c. 1937

THE BEQUEST
OF WILHELM KRUMBIEGEL
(1886–1960)

Since 2009, the Deutsche Fotothek of the Saxon State and University Library of Dresden
has stored a collection of works drawn from the career of carpenter and technical
draughtsman Wilhelm Krumbiegel. This collection includes roughly 100 photographs as
well as a portfolio of personal documents from Krumbiegel's estate. The images depict
the construction and furnishing of passenger ships built between 1910 and 1938.
Krumbiegel worked for Deutsche Werkstätten Hellerau from 1909 until the early 1950s,
first as a carpenter, then starting in 1915 as a draughtsman, and finally from the
mid 1920s on as a technician who also served as a site manager on larger projects.
He was involved in furnishing the express steamship *Bremen* and the cruise ship
Wilhelm Gustloff, among others. He also worked on the fit-out of the executives' floor in
the IG-Farben-Haus in Frankfurt am Main, designed by the architect Hans Poelzig.

(Marc Rohrmüller, Deutsche Fotothek)

INLAND NAVIGATION

ON LAKES AND RIVERS
LEIPZIG (1929)
ALLGÄU (1929)
DEUTSCHLAND (1935)

Deutschland on Lake Constance

Maiden voyage on 4 June 1935

ON LAKES AND RIVERS

At the beginning of the twentieth century, Germany experienced a boom not only in the number of ships traversing the high seas, but also in the production of passenger vessels operating on lakes and rivers. Pleasure cruises were popular, especially on the Rhine, Danube, Elbe and Moselle rivers as well as on Lake Constance. The companies offering these trips took great care to reproduce, in their inland ships, at least a portion of the cosmopolitan charm offered onboard the imposing luxury ocean liners of the day. They thereby sought to distance themselves from the stereotype of inland steamships as second-class 'floating pubs'. The new passenger ships were certainly not comparable to the 'ocean giants' in size, though, with regard to interiors, some of them were able to hold their own. Older ships were often modernised at great expense.

Harbour in Lindau

c. 1936

LEIPZIG (1929)

Sächsisch-Böhmische Dampfschiffahrt | Laubegast Shipyard (Dresden) | 70.1 m

The Sächsische Dampfschiffahrt company, founded in 1836, owns the world's largest and oldest fleet of paddle steamers, which are powered by two parallel rotating paddle wheels. The *Leipzig*, built in 1929 and capable of carrying up to 1,500 passengers, is the youngest and also the largest of Sächsische Dampfschiffahrt's fleet. She originally sailed as a 'saloon steamer' on the Elbe. During the Second World War, the *Leipzig* was converted into a hospital ship and painted camouflage grey. She carried the wounded out of Dresden following the devastating air raids in February 1945. A short time later she was badly damaged by an aerial bomb and had to undergo extensive repairs. As of 1947, the *Leipzig* once again sails the Elbe as a passenger ship, to the delight of her guests – including many day trippers and tourists.

Deutsche Werkstätten executed the interior furnishings for the public areas on the *Leipzig*, first class and second class alike. The plans were provided by the Munich architect Karl Bertsch, who designed the interiors of many ocean-going vessels as well.

Staircase

Waiter's work station

Public area

In 'tween decks

Leipzig

Moored on the Embankment

in Dresden, 1934

ALLGÄU (1929)

Deutsche Reichsbahn | Deggendorf Shipyard and Ironworks (Deggendorf) | 60.5 m

Following a general overhaul, the *Allgäu* returned to service as a German passenger ship on Lake Constance in 1949 and continued in this role for a further 50 years. Her interiors were deliberately designed to evoke the great luxury liners; this fact earned her the flattering nickname 'the *Bremen* of Lake Constance'.

All public areas in first class, along with the staircases, were fitted out by Deutsche Werkstätten. The smoking room was panelled in reddish-brown walnut, while the ladies' lounge was cladded in light red cherry wood. The large dining hall, in contrast, had flame birch wood panelling, which stood out against the blue velour carpet. The designs presumably came from Karl Bertsch.

The *Allgäu* was the first passenger ship on Lake Constance to be powered by a diesel motor. At 60.5 metres long – which was the maximum limit allowed – she could carry 1,200 passengers. Compared to the other, rather small boats on the lake, the modern double-decker, which was mainly used for special excursions, must have seemed like an ocean liner. When the Second World War began, she was decommissioned and docked in her home port of Lindau. After the war, she performed a stint as 'floating headquarters' for the French military police.

First-class area

With alcove seating

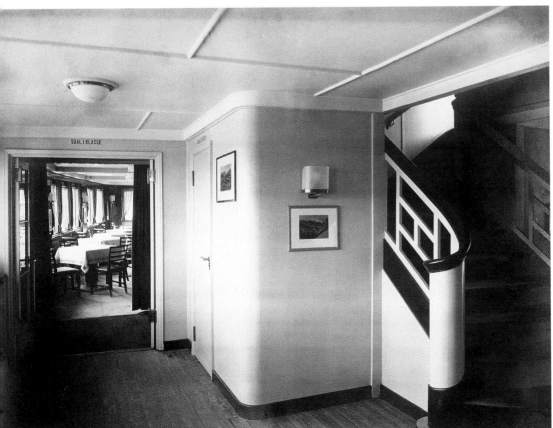

Staircase

View into the second-class dining hall

Staircase

Ascent to first class

Allgäu

Coming into harbour

c. 1935

DEUTSCHLAND (1935)

Deutsche Reichsbahn | Deggendorf Shipyard and Ironworks (Deggendorf) | 56.3 m

The *Deutschland* was a triple-deck motorship that belonged to the Bavarian fleet on Lake Constance. Following her launch in 1935, the *Deutschland* was used primarily for special excursions and prestigious occasions. Like the *Allgäu*, she was decommissioned during the Second World War and was later put to use by the French military, which renamed her *Rhin et Danube*. The occupying troops threw balls and dancing cruises on board, earning the ship the nickname

'Reine de la Nuit' (Queen of the Night). By 1949 the vessel was called the *Lindau* and was once again registered as a German passenger ship. There was one final name change, this time to *Überlingen*, along with a further 56 years of service on Lake Constance.

On the *Deutschland*, as on the *Allgäu*, Deutsche Werkstätten produced the interiors for the first-class passengers. These included an elegantly furnished smoking room, as well as an invitingly bright dining room.

First-class dining hall

With an eye-catching ceiling

construction

Deutschland

In her home port of Lindau, 1936

161

Staircase

Different perspectives

First-class smoking room

With rich wood panelling

HAMMER AND SICKLE

| REPARATIONS FOR THE SOVIET UNION |
| OLD ACQUAINTANCES |
| POBEDA (1952) |
| ERNST MAX JAHN |
| RUSS (1952) |

Asia

On the fitting-out quay

in Rostock-Warnemünde

1950

Admiral Nakhimov

c. 1955

REPARATIONS FOR THE SOVIET UNION

During the Second World War, passenger ships built by German shipping companies in most cases served as barracks and hospital ships, auxiliary cruisers and troop transporters. Many of these vessels were sunk during the course of the war, while others were rendered inoperable and left behind in shallow waters. The Third Reich capitulated in May of 1945 and was subsequently occupied by the victorious Allied Powers. The Soviet Union, in particular, demanded large reparation payments from Germany for the damage it had caused. Part of these 'payments' came in the form of rehabilitating former passenger steamships that had been captured by the Red Army or otherwise acquired by the Soviets. Almost all of the ships were in very bad shape and had to be towed to various shipyards for repairs, with most being brought to Rostock-Warnemünde or to Wismar. There the former 'ocean giants' were not only made seaworthy again and repainted; most of them were also provided with entirely new interiors.

Yuri Dolgorukiy

In the Warnow Shipyard in

Rostock-Warnemünde, 1959

OLD ACQUAINTANCES

Deutsche Werkstätten helped rehabilitate passenger ships on a large scale following the Second World War. Until recently, however, the details surrounding the projects were mostly unknown, even at Deutsche Werkstätten. A fascinating coincidence is that Deutsche Werkstätten had previously worked on at least two of the ships they were commissioned to overhaul – the *Magdalena* and the *Cordillera*. In the early 1950s, the Soviet Union had these vessels converted and refurbished. Renamed the *Pobeda* and the *Russ*, respectively, they sailed as passenger liners and cruise ships under their new flag. In addition, there is some evidence that Deutsche Werkstätten was also involved in rehabilitating the *Hamburg*. The original plan was to turn her back into a passenger steamer; however, she was ultimately repurposed as the *Yuri Dolgorukiy*, a mothership for whaling vessels.

Wreckage of the *Cordillera*

Before she was converted to the

passenger steamer *Russ*, c. 1948

POBEDA (1952)

Soviet Union | Wismar Repair Yard (Wismar) | 148.1 m

Originally built to service HAPAG's Central American line, this ship was first known as the *Magdalena*. From 1935 on, she operated under the name *Iberia*. Following the Second World War, the nearly 150-metre-long steamer was used by the British Royal Navy as temporary lodgings while docked in the Port of Kiel. This was appropriate, since she had served as a barracks ship during the final years of the war. Early in 1946, the *Iberia* was given to the Soviet Union as reparations. Under the new name *Pobeda* (Victory), she sailed for a time on the Black Sea, commuting between Crimea and the Caucasus. In 1950, she travelled to Wismar for a general overhaul.

On 8 May 1952, seven years to the day after the Third Reich's unconditional surrender, the completely revamped *Pobeda* was ceremonially turned over to its new captain. She went on to sail for another quarter of a century, working as a liner and cruise ship on the Black Sea.

Second-class dining hall

Different perspectives

During the *Pobeda*'s general overhaul in the early 1950s, Deutsche Werkstätten carried out the majority of the interior outfitting. This included the restaurant and music room in first class, the restaurant and dining hall in second class, and the luxury, state and officers' cabins. The Leipzig architect Ernst Max Jahn served as artistic director for the renovations. Room furnishings incorporated many valuable woods from all over the world, including East Indian rosewood, Honduran mahogany, Makassar ebony and ice birch.

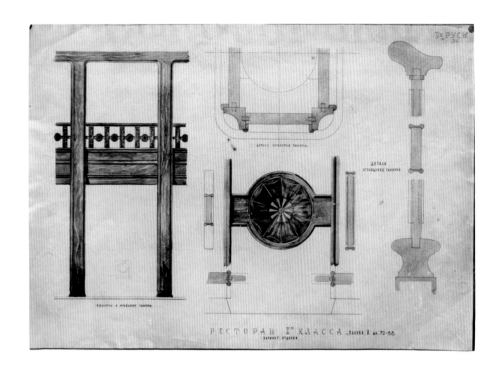

First-class restaurant

Design drawing of details and

wall elevation, c. 1950

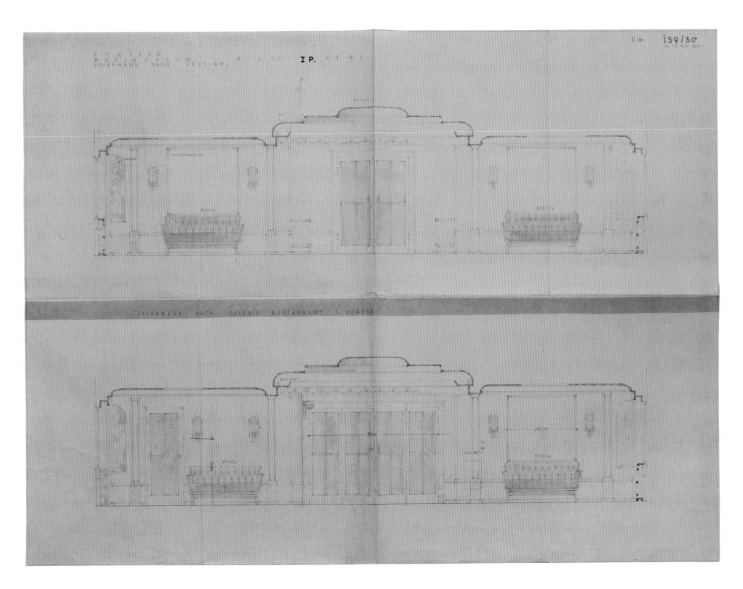

First-class music room

Wall elevation, c. 1950

Wall light

In the smoking room

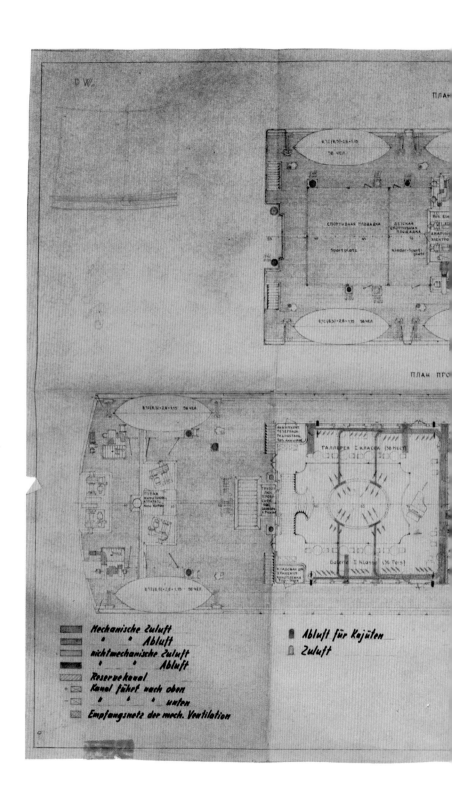

Ventilation plan

For the promenade, boat and

bridge decks, c. 1950

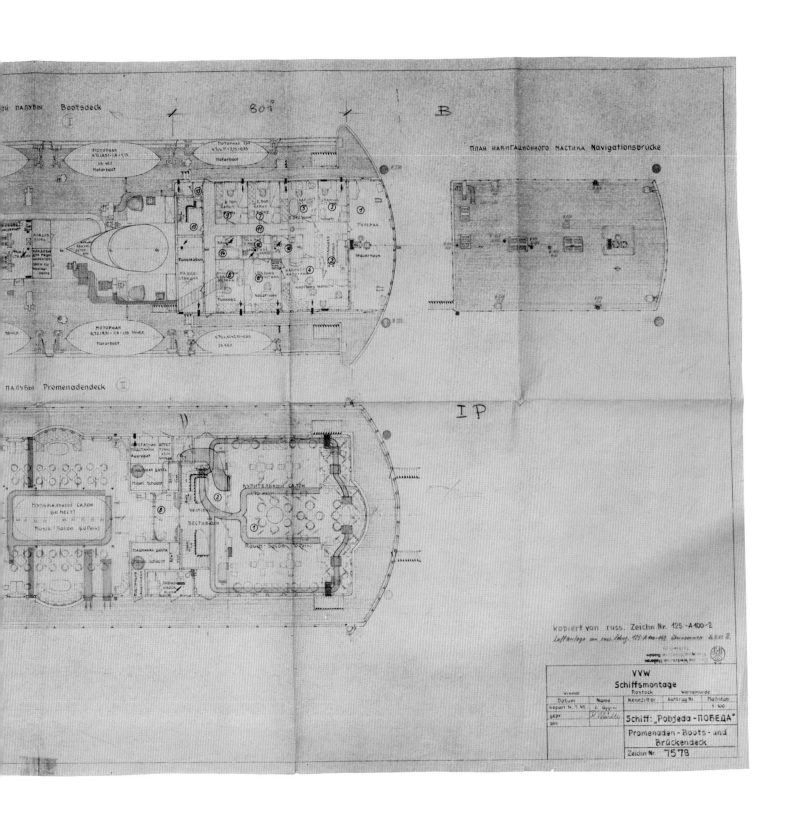

ий палубы Bootsdeck Ⓘ

80°

B

ПЛАН НАВИГАЦИОННОГО МОСТИКА Navigationsbrücke

Моторная
Motorboot

Моторная
Motorboot

Рулевая
Steuerhaus

Радио-
станция

Моторная
Motorboot

палубы Promenadendeck Ⓘ

IP

Музыкальный салон
Musik Salon (40 Pers.)

Курительный салон
Rauch-Salon (40 Pers.)

Vestibül
Вестибюль

kopiert von russ. Zeichn. Nr. 125-A100-2

VVW			
Schiffsmontage			
Wismar	Rostock	Warnemünde	
Datum	Name	Kennziffer	Auftrag Nr.	Maßstab
kopiert				1:100
gepr.		Schiff: „Pobjeda -ПОБЕДА"		
gez.		Promenaden- Boots- und		
		Brückendeck		
		Zeichn Nr. 7578		

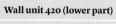

Wall unit 420 (lower part)

Oak, designed c. 1935

ERNST MAX JAHN

b. 1889 in Berlin | d. 1979 in Leipzig

Little is known about the life of Ernst Max Jahn, despite the fact that he was one of
Deutsche Werkstätten's most noteworthy designers, especially in the years after
the Second World War. Following his education as an artisan, he worked for the company
Carl Müller in Leipzig, where he received additional training as an interior designer.
He also took classes at the local school for decorative arts during this time. Jahn was
first commissioned by Deutsche Werkstätten as a freelancer in the early 1920s. In
the ensuing decades he produced numerous designs for furniture, with great success.
In addition, Jahn assisted Deutsche Werkstätten with several interior outfitting projects,
such as the German Democratic Republic's embassy in Prague (1954/1955). During
the renovation of the cruise ships *Pobeda* and *Russ* (1949–1951), he was tasked with
designing the public rooms in first class.

Bedroom 7791

c. 1955

RUSS (1952)

Soviet Union | Warnow Shipyard (Rostock-Warnemünde) | 159.8 m

The *Russ* was built in Hamburg by Blohm & Voss and was originally known as the *Cordillera*. She sailed the Central American line for HAPAG beginning in 1933. During the Second World War she served as a barracks ship, until an aerial bomb caused her to capsize near Świnoujście (in present-day Poland) in March of 1945. The protruding wreckage was salvaged after the war and was towed, at the order of the Soviet Union, to the Warnow Shipyard in Rostock-Warnemünde for a general overhaul. Renamed the *Russ*, this former HAPAG steamship served as a liner between Vladivostok and Petropawlowsk starting in 1952.

Deutsche Werkstätten had originally furnished the first-class dining hall on board the *Cordillera*, and for the ship's transformation into the *Russ* the company refurbished the same hall. In addition, Deutsche Werkstätten was also responsible for fitting out the lounge and music room in first class. The designs for these interiors came – as they did for the *Pobeda* – from Ernst Max Jahn.

First-class music room

Dance floor and grand piano

First-class dining hall

Mahogany panelling

First-class dining hall

Looking towards the staircase

EPILOGUE

Jan Jacobsen | Managing Director
Deutsche Werkstätten Hellerau GmbH

This is no mean achievement. After all, since our founding we have survived two world wars as well as the division and reunification of Germany.

So, I hope you will bear with me as I use this epilogue not only to address our current work in yacht outfitting, but also to draw a connection from our past to the future. Without understanding our tradition, it is quite difficult to understand our development.

Why Deutsche Werkstätten is still around

Museums connect us to our past. This book on ship interiors aims to do the same. Only we have never seen ourselves as a 'museum piece', but rather as a creative and innovative company that is at the cutting edge of its time.

When a curator from the Victoria & Albert Museum in London contacted us some time ago, requesting our support with the exhibition 'Ocean Liners: Speed and Style', she could hardly hide her surprise – and joy! – that Deutsche Werkstätten 'is still around', that we still exist and continue to innovate and impress, 120 years after the company was founded.

Since the very beginning Deutsche Werkstätten has been associated with visionary leadership. It was the courage and ingenuity of the company's founder, Karl Schmidt, that led him to convince the naval authorities in Kiel as well as the large German shipping companies of his vision for modern and elegant ship interiors at the dawn of the twentieth century.

Ninety years later, the company's new owner, Fritz Straub, built on this legacy of excellence and convinced the shipyards in Bremen and Kiel of the unyielding high quality of Deutsche Werkstätten's engineering and craftsmanship. As a result, we were entrusted with furnishing the interiors of privately owned superyachts and have now been active in this 'royal class' of interior outfitting for nearly twenty years.

The international clients who commission and own these yachts are, in a sense, the direct successors of the trans-atlantic passengers from the first half of the twentieth century. Their high expectations regarding quality, beauty, comfort and modernity inspire and drive us to combine the highest level of craftsmanship with state-of-the-art engineering. We think of our craft as a cultural asset, and are proud of making valuable things by hand here in Germany.

We work in a tightly knit niche market, where we know one another and treat each other with respect. Building these ships is the work of many hands and minds, everyone has their job and their specialisation. As an architect I love seeing designers, shipyard workers and our craftsmen collaborating to create a bespoke solution fulfilling the client's wish for a one-of-a-kind product.

In the end, each finished vessel reflects the personality of its owner. The journey to this point often lasts several years and requires intensive support and input. To fulfil this calling, we have built a strong and creative team that can bring form to vision and shoulder responsibilities. Anyone who has seen how our interiors are manufactured and then installed on board, or witnessed how much commitment and passion each of our craftsmen devotes to their tasks,

knows what I am talking about. This is where the spirit of our company can be most clearly perceived.

The outfitting of marine interiors is subject to requirements that differ considerably from those in residential construction. Technical know-how, innovative solutions and expertise are as essential as precision, perfection and the courage to try something new. We work true to the motto of our founder, Karl Schmidt – "Don't just do it differently, do it better" – keeping his spirit and attitude alive today.

And the future? We're not afraid of it – we're shaping it. One defining characteristic of Deutsche Werkstätten is its constant development: construction drawings in 3D, our own R&D division and a wide variety of special finishing techniques are just three examples of our undiminished ability to innovate.

What began 120 years ago as a combination of hand-craftsmanship and machine manufacturing continues today as a symbiosis of traditional skills and state-of-the-art production methods. Our value chain begins in the engineering department, where our resourceful design engineers work out solutions that are then realised by their colleagues in production. The products, either as single components or as assembly groups, are later installed with millimetre precision by our fitters on site. This is the only way to achieve the highest level of quality.

We will remain committed to yacht outfitting. However, providing interior furnishings for private jets or even luxury trains are the next logical steps in this development. Much of what we have learned from our work on yachts has already been applied to residential projects such as villas, chalets and apartments, where we have been responsible for the complete interior outfitting. It is also quite possible that we will return to designing and manufacturing our own furniture again. Building our brand, distinguishing ourselves from the competition and upholding our reputation for excellence remain our top priorities.

And by the way, we plan to keep growing. Next to our headquarters in Dresden-Hellerau we are planning a new campus that will include a new administrative building and an adjoining academy dedicated to training our employees. Perhaps even a small museum showcasing the historical development of our company.

As you can see, we are full of ideas to share with you as a client, designer, employee or visitor. You are invited to help us shape the future. I am confident you will sense 'the spirit of Hellerau' in your next encounter with Deutsche Werkstätten.

Yours,
Jan Jacobsen

MY A (2008)
Modern superyacht design

SHIP INDEX

Deutsche Werkstätten's contributions to the interiors for the following ships are verifiable. In addition, there is evidence of involvement on other projects, including the *Columbus* (1913), the *Cap Arcona* (1927), the *General Osorio* (1929) and the *Yuri Dolgorukiy* (1960).

*   Year given refers to the date of launch
     (Ship did not enter service under this name)

**   Year refers to completion of refurbishment

1904		### PRINZ ADALBERT
		Imperial Navy | Imperial Shipyard (Kiel)
		126.5 m | 6 070 GRT
		By Deutsche Werkstätten: Officers' mess (Design: Richard Riemerschmid)

1905		### BERLIN
		Imperial Navy | Imperial Shipyard (Gdansk)
		111.1 m
		By Deutsche Werkstätten: Officers' mess, captain's cabin (Design: Richard Riemerschmid)

1906		### ROON
		Imperial Navy | Imperial Shipyard (Kiel)
		127.8 m
		By Deutsche Werkstätten: Officers' mess, captain's cabin (Design: Richard Riemerschmid)

1907		### DANZIG
		Imperial Navy | Imperial Shipyard (Gdansk)
		111.1 m
		By Deutsche Werkstätten: Officers' mess, captain's cabin (Design: Richard Riemerschmid)

1907		### KRONPRINZESSIN CECILIE
		Norddeutscher Lloyd | AG Vulcan (Szczecin)
		215.3 m | 19 360 GRT
		By Deutsche Werkstätten: Imperial Suite (Design: Richard Riemerschmid)

0   50   100   150   200   250   300

1914*

J. H. BURCHARD

HAPAG | J. C. Tecklenborg (Geestemünde)
187.4 m | 19 980 GRT
By Deutsche Werkstätten: Staircase (Design: Karl Bertsch),
ladies' lounge (Design: Adelbert Niemeyer), suites
(Design: Richard Riemerschmid and Adelbert Niemeyer)

1922

THURINGIA

HAPAG | Howaldtswerke (Kiel)
150.9 m | 11 251 GRT
By Deutsche Werkstätten: Dining hall

1923

WESTPHALIA

HAPAG | Howaldtswerke (Kiel)
150.9 m | 11 253 GRT
By Deutsche Werkstätten: Staircases

1924

DEUTSCHLAND

HAPAG | Blohm & Voss (Hamburg)
191.2 m | 20 602 GRT
By Deutsche Werkstätten: Stateroom (Design: Adelbert
Niemeyer and Bruno Paul), ladies' writing room
(Design: Adelbert Niemeyer), second-class staircase

1924

NJASSA

HAPAG | Blohm & Voss (Hamburg)
132.1 m | 8754 GRT
By Deutsche Werkstätten: First-class dining hall,
first-class ladies' lounge

| 0 | 50 | 100 | 150 | 200 | 250 | 300 |

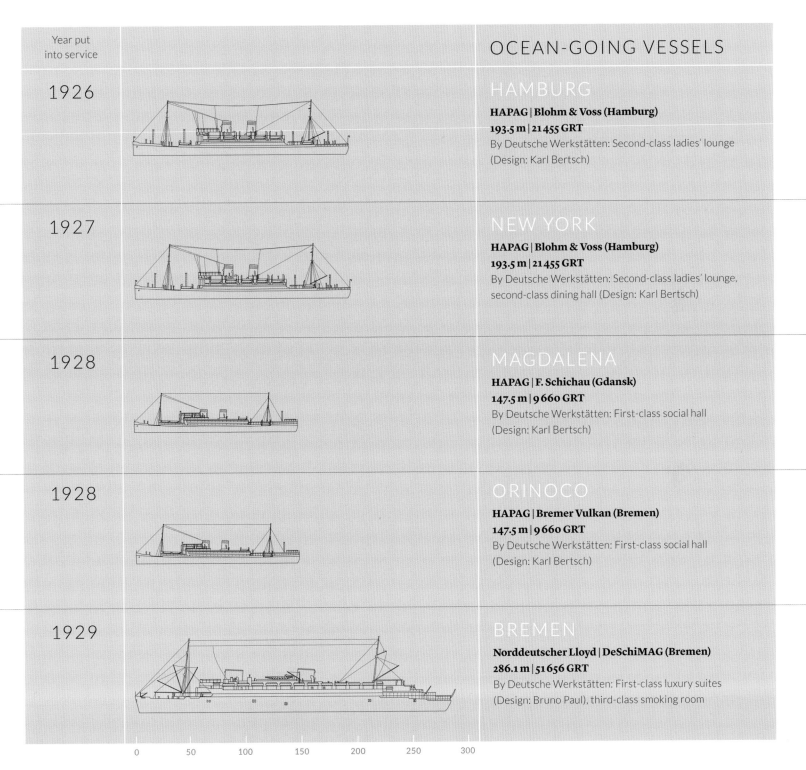

Year put into service		OCEAN-GOING VESSELS
1926		## HAMBURG **HAPAG | Blohm & Voss (Hamburg)** **193.5 m | 21 455 GRT** By Deutsche Werkstätten: Second-class ladies' lounge (Design: Karl Bertsch)
1927		## NEW YORK **HAPAG | Blohm & Voss (Hamburg)** **193.5 m | 21 455 GRT** By Deutsche Werkstätten: Second-class ladies' lounge, second-class dining hall (Design: Karl Bertsch)
1928		## MAGDALENA **HAPAG | F. Schichau (Gdansk)** **147.5 m | 9 660 GRT** By Deutsche Werkstätten: First-class social hall (Design: Karl Bertsch)
1928		## ORINOCO **HAPAG | Bremer Vulkan (Bremen)** **147.5 m | 9 660 GRT** By Deutsche Werkstätten: First-class social hall (Design: Karl Bertsch)
1929		## BREMEN **Norddeutscher Lloyd | DeSchiMAG (Bremen)** **286.1 m | 51 656 GRT** By Deutsche Werkstätten: First-class luxury suites (Design: Bruno Paul), third-class smoking room

0   50   100   150   200   250   300

OCEAN-GOING VESSELS

1929

MILWAUKEE

HAPAG | Blohm & Voss (Hamburg)
175.5 m | 16 699 GRT
By Deutsche Werkstätten: Third-class smoking room
(Design: Karl Bertsch)

1929

ST. LOUIS

HAPAG | Bremer Vulkan (Bremen)
174.9 m | 16 732 GRT
By Deutsche Werkstätten: Third-class smoking room
(Design: Karl Bertsch)

1933

CARIBIA

HAPAG | Blohm & Voss (Hamburg)
159.9 m | 12 049 GRT
By Deutsche Werkstätten: First-class dining hall
(Design: Karl Bertsch)

1933

CORDILLERA

HAPAG | Blohm & Voss (Hamburg)
159.8 m | 12 055 GRT
By Deutsche Werkstätten: First-class dining hall
(Design: Karl Bertsch)

1938

BOISSEVAIN

Koninklijke Paketvaart Maatschappij
Blohm & Voss (Hamburg) | 170.5 m | 14 134 GRT
By Deutsche Werkstätten: Dining hall, lounge,
foyer, staircases (Design: Bruno Paul)

0   50   100   150   200   250   300

OCEAN-GOING VESSELS

1938

WILHELM GUSTLOFF

Deutsche Arbeitsfront | Blohm & Voss (Hamburg)
208.5 m | 25 484 GRT
By Deutsche Werkstätten: Great Hall
(Design: Woldemar Brinkmann)

1939

ROBERT LEY

Deutsche Arbeitsfront | Howaldtswerke (Hamburg)
203.8 m | 27 288 GRT
By Deutsche Werkstätten: Sunroom, sports hall

1950

ASIA

Soviet Union | Warnow Shipyard (Rostock-
Warnemünde) | 149.5 m | 11 453 GRT
By Deutsche Werkstätten: Second-class social hall

1952**

POBEDA

Soviet Union | Wismar Repair Yard (Wismar)
148.1 m | 9 829 GRT
By Deutsche Werkstätten: First-class restaurant,
second-class restaurant and dining hall, first-class music
room, various cabins (Design: Ernst Max Jahn)

1952

RUSS

Soviet Union | Warnow Shipyard (Rostock-
Warnemünde) | 159.8 m | 12 931 GRT
By Deutsche Werkstätten: First-class dining hall,
first-class music room and lounge
(Design: Ernst Max Jahn)

| 0 | 50 | 100 | 150 | 200 | 250 | 300 |

OCEAN-GOING VESSELS

1953

ALEXANDER MOZHAYSKY

Soviet Union | Mathias-Thesen-Werft (Wismar)
152.4 m | 9 922 GRT
By Deutsche Werkstätten: First-class smoking room,
first-class music room

1957

ADMIRAL NAKHIMOV

Soviet Union | Warnow Shipyard (Rostock-
Warnemünde) | 174.3 m | 17 053 GRT
By Deutsche Werkstätten: First-class dining hall

INLAND VESSELS

1929

LEIPZIG

Sächsisch-Böhmische Dampfschiffahrt
Laubegast Shipyard (Dresden) | 70.1 m
By Deutsche Werkstätten: Public rooms, staircase
(Design: Karl Bertsch)

1929

ALLGÄU

Deutsche Reichsbahn | Deggendorf Shipyard and
Ironworks (Deggendorf) | 60.5 m
By Deutsche Werkstätten: First-class public rooms,
staircase

1935

DEUTSCHLAND

Deutsche Reichsbahn | Deggendorf Shipyard and
Ironworks (Deggendorf) | 56.5 m
By Deutsche Werkstätten: First-class public rooms,
staircase

| 0 | 50 | 100 | 150 | 200 | 250 | 300 |

NAMES

SOURCES

Archives

- Architekturmuseum der Technischen Universität München
- Sächsisches Staatsarchiv – Hauptstaatsarchiv Dresden
- SLUB – Deutsche Fotothek
- Schiffbau- und Schifffahrtsmuseum Rostock

Journals

- Dekorative Kunst
- Innendekoration
- Kunst und Handwerk
- Kunstgewerbeblatt
- Werft, Reederei, Hafen

Literature (Selection)

- Arnold, Klaus-Peter, *Vom Sofakissen zum Städtebau: Die Geschichte der Deutschen Werkstätten und der Gartenstadt Hellerau*, Dresden/Basel: Verlag der Kunst, 1993.
- Finamore, Daniel and Ghislaine Wood (ed.), *Ocean Liners*, London: V&A, 2018.
- Fritz, Karl F. and Reiner Jäckle, *Der Siegeszug der Motorschiffe auf dem Bodensee*, Erfurt: Sutton, 2015.
- Kludas, Arnold, *Die Geschichte der deutschen Passagierschiffahrt* (5 vols.), Augsburg: Weltbild, 1994.
- Kludas, Arnold, *Die Geschichte der Hapag-Schiffe* (5 vols.), Bremen: Hauschild, 2007–2010.
- Müller, Frank and Wolfgang Quinger, *Die Dresdner Raddampferflotte*, Bielefeld: Delius Klasing, 2007.
- Nerdinger, Winfried (ed.), Richard Riemerschmid, *Vom Jugendstil zum Werkbund: Werke und Dokumente*, Munich: Prestel, 1982.
- Rothe, Claus, *Deutsche Ozeanpassagierschiffe: 1896 bis 1918*, Berlin (East): Transpress, 1986.
- Rothe, Claus, *Deutsche Ozeanpassagierschiffe: 1919 bis 1985*, Berlin (East): Transpress, 1987.
- Schwerdtner, Nils, *German Luxury Ocean Liners: From Kaiser Wilhelm der Grosse to AIDAstella*, Gloucestershire: Amberley, 2013.
- Thiel, Reinhold, *Die Geschichte des Norddeutschen Lloyd: 1857–1970* (5 vols.), Bremen: Hauschild, 1999.
- Trennheuser, Matthias, *Die innenarchitektonische Ausstattung deutscher Passagierschiffe zwischen 1880 und 1940*, Bremen: Hauschild, 2010.
- Wiborg, Susanne and Klaus Wiborg, *1847–1997 Unser Feld ist die Welt: 150 Jahre Hapag-Lloyd*, Hamburg: Springer, 1997.
- Wichmann, Hans, *Aufbruch zum neuen Wohnen: Deutsche Werkstätten und WK-Verband 1898–1990*, Munich: Prestel, 1992.
- Wilson, Edward A., *Soviet Passenger Ships 1917–1977*, Kendal: World Ship Society, 1978.
- Witthöft, Hans Jürgen, *Gebaut bei Blohm + Voss*, Hamburg: Koehler, 2004.
- Ziffer, Alfred (ed.), *Bruno Paul: Deutsche Raumkunst und Architektur zwischen Jugendstil und Moderne*, Munich: Klinkhardt & Biermann, 1992.

IMAGES

Kunstgewerbemuseum, Staatliche Kunstsammlungen Dresden,
Foto: Robert Vanis
pp. 27, 28

Library of Congress, LC-DIG-ppmsca-02202
pp. 14/15

Matthias Lüdecke
p. 10

Münchner Stadtmuseum, Sammlung Angewandte Kunst
p. 67 l

Münchner Stadtmuseum, Sammlung Graphik/Gemälde
p. 91

NH 64262 courtesy of the Naval History & Heritage Command
pp. 48/49

Nürnberg, Germanisches Nationalmuseum, Deutsches
Kunstarchiv, NL Riemerschmid, Richard, I,A-1 (0030)
p. 53

Rue des Archives / Collection Grégoire /
Süddeutsche Zeitung Photo
p. 110

RVM, Bildarchiv Eisenbahnstiftung
p. 149

Sächsisches Staatsarchiv – Hauptstaatsarchiv Dresden
pp. 4 b, 6 br, 8 b, 16, 19, 24, 26/27, 29, 30/31, 32/33, 38/39,
40, 47, 54/55, 65, 67 r, 72, 75 r, 76/77, 83 t, 83 bl, 83 br, 84/85,
86, 87 t, 87 b, 90 t, 98/99, 99, 100, 102 t, 103, 104/105, 105,
115 b, 116, 117, 121, 135, 136 t, 137, 145, 151, 153, 155 t,
155 b, 156/157, 159, 162 t, 162 b, 163, 170/171, 171, 172 t,
172 b, 173, 174, 174/175, 177

Sammlung Lemachko
pp. 169, 178

Sammlung Matthias Trennheuser
p. 111 t

Scherl / Süddeutsche Zeitung Photo
pp. 6 t, 78/79, 108

Slg. A. Heer
pp. 148, 157, 160/161

Slg. A. Heer / Schöning Verlag, Lübeck
p. 158

SLUB / Deutsche Fotothek
pp. 9 t, 20, 62/63, 92/93, 94/95, 100/101, 109, 112/113,
114, 115 t, 120, 125 r, 128, 129, 130/131, 131, 132, 132/133,
134, 138, 139, 140, 140/141, 141, 144 b, 164/165

SLUB / Deutsche Fotothek, Franz Grasser
pp. 76, 126/127

SLUB / Deutsche Fotothek, Franz Stoedtner
pp. 5 t, 58/59

SLUB / Deutsche Fotothek, Oswald Lübeck
p. 17

SLUB / Deutsche Fotothek, Photographisches
Atelier der Hamburg-Amerika-Linie
pp. 106/107, 107

SLUB / Deutsche Fotothek, Walter Hahn
pp. 8 t, 146/147, 150

SLUB / Deutsche Fotothek, Walter Möbius
p. 152/153

IMPRINT

© 2018
Sandstein Verlag, Dresden, and publisher

Publisher
Deutsche Werkstätten

Managing Editor
Konstantin Kleinichen, Deutsche Werkstätten

Picture Editor
Steffen Jungmann

Translation
Jonathan Pattishall

Editor
Christine Jäger-Ulbricht, Sandstein Verlag
Adele Arber, Deutsche Werkstätten

Layout
Michaela Klaus, Sandstein Verlag

Cover Design
Sandra Püschel, Deutsche Werkstätten

Print and Reprographics
Katharina Stark, Jana Neumann, Sandstein Verlag

Printing and Processing
FINIDR s. r. o., Český Těšín

The German National Library has registered this publication in the Deutsche Nationalbibliografie; detailed bibliographical information can be accessed at the following Web site: http://dnb.dnb.de.

www.sandstein-verlag.de
ISBN 978-3-95498-445-9